AF818440

Table of Contents

The AI Strategy Planning Kit

Your Comprehensive Guide to Implementing AI in Business

In today's rapidly evolving business landscape, artificial intelligence (AI) stands out as a pivotal technology driving innovation, efficiency, and competitive advantage. "The AI Strategy and Implementation Planning Kit" is a meticulously crafted guide designed to help businesses of all sizes harness the transformative power of AI. This book provides comprehensive, step-by-step instructions, practical worksheets, and real-world case studies to ensure successful AI implementation across various industries, including retail, healthcare, manufacturing, and finance.

Key features of this guide include:

- **AI Maturity Assessment:** Tools and questionnaires to evaluate your organization's readiness for AI.
- **Data Infrastructure:** Strategies for data collection, storage, accessibility, and integration.
- **Technology Stack:** Insights into hardware, software, and network infrastructure needed for AI.
- **Skill Set Development:** Training programs, hiring strategies, and team collaboration tips for building a robust AI talent pool.
- **Business Alignment:** Methods to align AI initiatives with strategic business objectives, manage risks, and ensure leadership support.
- **Implementation Roadmaps:** Detailed 6-month and 12-month roadmaps to guide your AI projects from conception to execution.
- **Case Studies:** Real-world examples of AI applications in various sectors, showcasing successful outcomes and lessons learned.

By following the guidance provided in this book, businesses can navigate the complexities of AI integration, achieve measurable improvements, and stay ahead in an increasingly competitive market.

Introduction: Embracing the Future with AI

Artificial intelligence (AI) is no longer a futuristic concept reserved for tech giants and academic researchers. It has become a practical and indispensable tool for businesses across the globe, driving innovation, enhancing efficiency, and creating new avenues for growth. As AI continues to evolve, its applications in business are expanding, offering unprecedented opportunities for those who are ready to embrace its potential.

However, the journey to successful AI implementation is fraught with challenges. Many organizations struggle with understanding where to start, how to integrate AI into their existing operations, and how to measure its impact. This is where the **AI Strategy and Implementation Planning Kit** comes into play.

This book is designed to demystify AI for business leaders, managers, and IT professionals. It offers a comprehensive guide to planning and executing AI initiatives, tailored to the unique needs and goals of your organization. Whether you are just beginning to explore AI or looking to enhance your current AI capabilities, this book provides the tools, strategies, and insights you need to succeed.

Why AI Matters for Your Business

AI has the power to transform businesses by automating routine tasks, uncovering hidden insights from data, and enabling smarter decision-making. From improving customer experiences to optimizing supply chains and enhancing product development, the applications of AI are vast and varied. By integrating AI into your business strategy, you can achieve significant cost savings, boost productivity, and gain a competitive edge in your industry.

What You Will Learn

The **AI Strategy and Implementation Planning Kit** is structured to guide you through every step of the AI implementation process:

- **Assessing AI Readiness:** Understand your organization's current capabilities and identify areas for improvement.
- **Building a Strong Data Foundation:** Learn how to collect, store, and manage data effectively to support AI initiatives.

- **Choosing the Right Technology Stack:** Discover the hardware, software, and network infrastructure needed for AI.
- **Developing AI Skills and Teams:** Find out how to train your staff, hire AI talent, and foster team collaboration.
- **Aligning AI with Business Goals:** Ensure your AI projects support your strategic objectives and deliver tangible benefits.
- **Executing AI Projects:** Follow detailed roadmaps and checklists to implement AI solutions successfully.
- **Learning from Real-World Examples:** Gain insights from case studies across various industries.

Our Commitment

Our goal is to empower you with the knowledge and tools to navigate the complexities of AI and unlock its full potential for your business. As you embark on this journey, remember that AI is not just about technology; it's about transforming your business processes, culture, and mindset. With the right strategy and approach, you can lead your organization into a future where AI drives sustainable growth and innovation.

Welcome to the **AI Strategy and Implementation Planning Kit.** Let's begin this transformative journey together.

STEPS TO FOLLOW

1. The **AI Maturity Assessment** is a detailed questionnaire to evaluate your organization's current capabilities and readiness for AI implementation. **Complete the AI Maturity Assessment Questionnaire** by dividing the questionnaire up among your team members who are predominately responsible for the following business areas:

   Sections included: (Fill in the responsible person's name or team name)
   - **Data Infrastructure**: Availability and quality of data.
     - PERSON RESPONSIBLE ______________________________
   - **Technology Stack**: Existing hardware and software capabilities.
     - PERSON RESPONSIBLE ______________________________
   - **Skill Set**: In-house talent and expertise.
     - PERSON RESPONSIBLE ______________________________
   - **Business Alignment**: Alignment of AI initiatives with business goals.
     - PERSON RESPONSIBLE ______________________________

Ensure team members answer each question as deeply and knowingly as possible.

2. Compile all completed questions and answers together into one PDF file. This will become a vital resource for all team members to refer back to for completing the Implementation Checklists.

3. Choose either the **6-month or 12-month Roadmap**. Fill in the name of the AI Project and a specific description, i.e., "*Incorporate AI to manage inventory better and reduce stockouts by 30% and increase sales by 15%.*"

4. Using the **Implementation Checklists,** keep track of the progress of each phase of the AI Project. Systematically move through each checklist based on the **6-month or 12-month Roadmap** you have chosen.

5. Finally, using the **ROI (Return on Investment) AI Project Worksheet**, determine whether the AI Project will have a positive return on the investment.

AI Maturity Assessment Questionnaire

Data Infrastructure: Data Collection

Businesses collect a wide variety of data to support their operations, make informed decisions and enhance their strategies. Here are some examples of different types of data a business would typically collect:

1. **Customer Data:**
   - Contact Information: Names, addresses, phone numbers, email addresses.
   - Demographic Information: Age, gender, occupation, income level.
   - Purchase History: Products or services bought, frequency of purchases.
   - Feedback and Reviews: Customer satisfaction surveys, product reviews, complaints.
   - Loyalty Program Data: Points earned and redeemed, membership status.
2. **Sales Data:**
   - Sales Transactions: Dates, amounts, payment methods.
   - Sales Trends: Seasonal trends, peak sales periods.
   - Sales Performance: Sales by product, sales by region, sales by salesperson.
3. **Financial Data:**
   - Revenue and Expenses: Income statements, balance sheets, cash flow statements.
   - Budgeting Data: Projected budgets, actual spending, variances.
   - Tax Information: Tax returns, tax payments, tax deductions.
4. **Operational Data:**
   - Inventory Levels: Stock quantities, reorder points, stock turnover rates.

- Supply Chain Information: Supplier details, delivery times, shipment tracking.
- Production Data: Manufacturing schedules, production costs, quality control metrics.

5. **Marketing Data:**
   - Campaign Performance: Click-through rates, conversion rates, return on investment (ROI).
   - Market Research: Competitor analysis, market trends, customer preferences.
   - Advertising Data: Ad impressions, ad spend, engagement metrics.
6. **Employee Data:**
   - Personal Information: Names, addresses, contact details.
   - Employment Records: Job titles, employment dates, performance evaluations.
   - Payroll Data: Salaries, wages, bonuses, deductions.
7. **Website and Digital Analytics:**
   - Traffic Data: Number of visitors, page views, bounce rates.
   - User Behavior: Time spent on site, navigation paths, interaction patterns.
   - E-commerce Data: Online sales, shopping cart abandonment rates, payment methods.
8. **Compliance and Legal Data:**
   - Regulatory Compliance: Documentation of compliance with industry regulations.
   - Legal Contracts: Contracts with customers, suppliers, employees.
   - Incident Reports: Records of any legal issues or compliance breaches.
9. **Customer Service Data:**
   - Support Tickets: Number of tickets, resolution times, customer satisfaction ratings.
   - Call Logs: Duration of calls, issues addressed, follow-up actions.
   - Chat Transcripts: Conversations from live chat support, common inquiries.
10. **Product Data:**
    - Product Specifications: Details about features, components, materials.
    - Product Performance: Reliability data, failure rates, customer returns.
    - Development Data: Research and development costs, time to market.

These data types help businesses understand their customers, optimize operations, improve products and services, and maintain financial health.

What types of data does your business currently collect?

The frequency of data collection in a business can vary depending on the type of data and its intended use. Here are some examples of different data types and their typical collection frequencies:

1. **Customer Data:**
   - **Contact Information:** Collected at the time of initial interaction or purchase, updated as needed.
   - **Demographic Information:** Collected periodically through surveys or at the time of initial interaction.
   - **Purchase History:** Collected in real-time with each transaction.
   - **Feedback and Reviews:** Collected continuously through feedback forms, surveys, and review platforms.
   - **Loyalty Program Data:** Collected in real-time with each interaction within the program.
2. **Sales Data:**
   - **Sales Transactions:** Collected in real-time with each transaction.
   - **Sales Trends:** Analyzed monthly, quarterly, and annually.
   - **Sales Performance:** Reviewed weekly, monthly, and quarterly.
3. **Financial Data:**
   - **Revenue and Expenses:** Collected daily, weekly, and monthly.
   - **Budgeting Data:** Collected monthly and reviewed quarterly.
   - **Tax Information:** Collected quarterly and annually.
4. **Operational Data:**
   - **Inventory Levels:** Monitored daily and updated in real-time.
   - **Supply Chain Information:** Collected continuously as shipments and deliveries occur.
   - **Production Data:** Collected daily and reviewed weekly.
5. **Marketing Data:**
   - **Campaign Performance:** Collected daily or weekly during the campaign.
   - **Market Research:** Conducted periodically, often quarterly or annually.
   - **Advertising Data:** Collected in real-time and reviewed weekly or monthly.
6. **Employee Data:**
   - **Personal Information:** Collected at the time of hiring and updated as needed.
   - **Employment Records:** Updated as events occur (e.g., promotions, evaluations).
   - **Payroll Data:** Collected bi-weekly or monthly.
7. **Website and Digital Analytics:**
   - **Traffic Data:** Collected in real-time and reviewed daily, weekly, and monthly.
   - **User Behavior:** Collected in real-time and analyzed weekly.
   - **E-commerce Data:** Collected in real-time and reviewed daily.
8. **Compliance and Legal Data:**
   - **Regulatory Compliance:** Collected as needed, typically annually or as regulations change.
   - **Legal Contracts:** Collected and updated as new contracts are signed.
   - **Incident Reports:** Collected as incidents occur.
9. **Customer Service Data:**
   - **Support Tickets:** Collected in real-time with each support interaction.

- **Call Logs:** Collected in real-time and reviewed daily or weekly.
- **Chat Transcripts:** Collected in real-time and reviewed periodically.

10. **Product Data:**
    - **Product Specifications:** Collected during product development and updated as changes occur.
    - **Product Performance:** Collected continuously from customer feedback and returns.
    - **Development Data:** Collected during the development cycle and reviewed at key milestones.

These frequencies ensure that businesses have up-to-date and relevant information to make informed decisions and respond promptly to changes in the market or operational conditions.

How frequently is data collected and updated?

Gaps in data collection can arise for various reasons, and they can impact the quality and reliability of the data. Here are some common types of gaps that could occur:

1. **Incomplete Data:**
    - **Missing Entries:** Some data points may not be recorded, such as missing customer contact information or incomplete survey responses.
    - **Unfilled Fields:** Fields in a database may be left blank due to oversight or lack of information.
2. **Inconsistent Data:**
    - **Different Formats:** Data collected in varying formats can cause inconsistencies, such as different date formats or units of measurement.
    - **Varying Quality:** Data collected from multiple sources may differ in quality, leading to unreliable or conflicting information.
3. **Outdated Data:**
    - **Stale Data:** Data that is not updated regularly can become outdated, such as old contact details or obsolete financial information.
    - **Lag in Data Entry:** Delays in entering data can result in using outdated information for decision-making.
4. **Biased Data:**
    - **Sampling Bias:** Data collected from a non-representative sample can lead to biased conclusions, such as surveying only a specific customer segment.

- **Self-Selection Bias:** When participants choose to provide data voluntarily, it may not represent the entire population accurately.

5. **Lack of Integration:**
   - **Siloed Data:** Data stored in separate systems or departments may not be integrated, preventing a comprehensive view of the information.
   - **Incompatible Systems:** Different software or systems that cannot communicate with each other can lead to data gaps.
6. **Data Entry Errors:**
   - **Human Error:** Mistakes made during manual data entry, such as typographical errors or incorrect values.
   - **Automated Errors:** Errors in automated data collection processes, such as incorrect sensor readings or software bugs.
7. **Insufficient Data Collection:**
   - **Limited Scope:** Collecting only a narrow range of data can result in missing important information, such as only tracking sales without considering customer feedback.
   - **Frequency Issues:** Data not collected frequently enough to capture important changes or trends, such as only annual financial reviews in a volatile market.
8. **Technical Issues:**
   - **System Failures:** Data loss due to hardware or software failures, such as server crashes or database corruption.
   - **Connectivity Issues:** Interruptions in data collection due to network problems or lack of internet access.
9. **Privacy and Compliance Constraints:**
   - **Regulatory Restrictions:** Legal constraints on the type of data that can be collected, such as GDPR or HIPAA regulations.
   - **Consent Issues:** Lack of proper consent from individuals can limit the data that can be collected.
10. **Environmental Factors:**
    - **External Influences:** Events such as natural disasters, economic downturns, or political instability can disrupt data collection efforts.

Identifying and addressing these gaps is crucial for ensuring the accuracy, reliability, and usefulness of the data collected by a business.

Are there any gaps in the data collection process? If so, please describe them.

Data Infrastructure: Data Storage

Business data can be stored in various ways depending on the type, volume, and usage requirements of the data. Here are some common methods of data storage used by businesses:

1. **Databases:**
   - **Relational Databases:** Store data in structured tables with rows and columns, using SQL for querying. Examples include MySQL, PostgreSQL, and Oracle.
   - **NoSQL Databases:** Designed for unstructured or semi-structured data, offering flexibility and scalability. Examples include MongoDB, Cassandra, and Redis.
2. **Data Warehouses:**
   - **Enterprise Data Warehouses:** Central repositories for storing large volumes of structured data from multiple sources, optimized for querying and analysis. Examples include Amazon Redshift, Google BigQuery, and Snowflake.
   - **Data Marts:** Subsets of data warehouses tailored to specific business lines or departments for targeted analysis.
3. **Cloud Storage:**
   - **Public Cloud Storage:** Services provided by third-party vendors, offering scalable and flexible storage solutions. Examples include Amazon S3, Microsoft Azure Blob Storage, and Google Cloud Storage.
   - **Private Cloud Storage:** Cloud infrastructure dedicated to a single organization, offering enhanced control and security.
4. **On-Premises Storage:**
   - **File Servers:** Centralized servers that store and manage files, accessible over a network.
   - **Network-Attached Storage (NAS):** Dedicated storage devices that provide file-based storage services to other devices on a network.

- **Storage Area Networks (SAN):** High-speed networks that provide block-level storage to servers, typically used for mission-critical applications.

5. **Data Lakes:**
   - **Data Lakes:** Large repositories that store raw data in its native format, allowing for flexible analysis and processing. Examples include Hadoop Distributed File System (HDFS) and Amazon S3-based data lakes.
6. **Hybrid Storage:**
   - **Hybrid Solutions:** Combining on-premises and cloud storage to balance performance, cost, and security requirements. Examples include hybrid cloud storage solutions like Azure Stack and AWS Outposts.
7. **Backup and Archival Storage:**
   - **Backup Solutions:** Regularly scheduled backups of data to protect against data loss, stored on different media such as external hard drives, cloud storage, or tape drives.
   - **Archival Storage:** Long-term storage solutions for preserving historical data, often on cost-effective and durable media like optical disks or tape.
8. **Data Management Platforms:**
   - **Data Management Systems:** Integrated platforms that provide tools for data integration, cleansing, and governance. Examples include Informatica, Talend, and IBM InfoSphere.
9. **Document Management Systems:**
   - **Document Repositories:** Systems designed for storing, managing, and tracking electronic documents and images of paper-based information. Examples include SharePoint, DocuWare, and M-Files.
10. **Content Management Systems (CMS):**
    - **Web Content Management:** Platforms for managing digital content, especially for websites and online applications. Examples include WordPress, Drupal, and Joomla.
11. **Blockchain Storage:**
    - **Blockchain Technology:** Distributed ledger technology for secure and immutable data storage, often used for transactions, contracts, and records. Examples include Ethereum and Hyperledger.
12. **Edge Storage:**
    - **Edge Computing:** Storage solutions located closer to data sources or end-users to reduce latency and improve performance, often used in IoT applications.

How is your data stored? E.g., cloud, on-premises, hybrid?

Data cleaning and validation are crucial steps in ensuring the quality and reliability of data before analysis or use in business processes. Here are some common procedures for data cleaning and validation:

Data Cleaning Procedures

1. **Removing Duplicates:**
    - Identify and remove duplicate records to ensure each entry is unique.
2. **Handling Missing Values:**
    - **Imputation:** Fill in missing values using statistical methods such as mean, median, or mode imputation.
    - **Deletion:** Remove records with missing values if they are not significant or if imputation is not feasible.
    - **Flagging:** Mark missing values for special treatment in analysis.
3. **Standardizing Formats:**
    - Ensure consistency in data formats, such as date formats (e.g., YYYY-MM-DD), phone numbers, and addresses.
    - Normalize text data to a consistent case (e.g., all lowercase) or remove special characters.
4. **Correcting Errors:**
    - Identify and correct typographical errors, data entry mistakes, and inconsistencies.
    - Use algorithms to detect outliers and anomalies that may indicate errors.
5. **Handling Outliers:**
    - **Removal:** Exclude outliers from the dataset if they result from data entry errors.
    - **Transformation:** Apply transformations to reduce the impact of outliers on analysis.
6. **Data Enrichment:**
    - Enhance data quality by integrating additional information from external sources or reference data.
7. **Normalization:**
    - Adjust values measured on different scales to a common scale, often used for numerical data.
8. **Splitting and Merging:**
    - Split combined data fields (e.g., full names into first and last names) or merge related fields to form a composite field.

Data Validation Procedures

1. **Schema Validation:**
    - Ensure data conforms to a predefined schema, including data types, field lengths, and mandatory fields.
2. **Range Checks:**
    - Validate numerical data to ensure values fall within a specified range (e.g., age between 0 and 120).

3. **Format Checks:**
    - Verify that data follows a specific format or pattern, such as email addresses, phone numbers, or postal codes.
4. **Consistency Checks:**
    - Ensure related data fields are consistent with each other (e.g., start date is before end date).
5. **Uniqueness Checks:**
    - Confirm that unique fields (e.g., IDs, social security numbers) do not contain duplicate values.
6. **Referential Integrity Checks:**
    - Validate relationships between tables, such as ensuring foreign keys match primary keys in related tables.
7. **Logic Checks:**
    - Apply business rules to validate data logic (e.g., an order date should not be in the future).
8. **Cross-Field Validation:**
    - Check for consistency across multiple fields (e.g., total amount should equal the sum of individual line items).
9. **Audit Trails:**
    - Maintain logs of data changes and validations performed to ensure transparency and traceability.
10. **Automated Validation:**
    - Implement automated validation scripts and tools to continuously monitor data quality.

Tools and Techniques

1. **ETL Tools:**
    - Extract, Transform, Load (ETL) tools like Informatica, Talend, and Microsoft SSIS provide built-in data cleaning and validation capabilities.
2. **Data Quality Tools:**
    - Specialized tools like Trifacta, DataCleaner, and OpenRefine offer advanced data cleaning and profiling features.
3. **Scripting Languages:**
    - Use programming languages like Python (with libraries such as Pandas and NumPy) or R for custom data cleaning and validation scripts.
4. **Database Functions:**
    - SQL queries and functions for data validation and cleaning within relational databases.
5. **Data Profiling:**
    - Analyzing data to understand its structure, content, and quality to identify issues.

By applying these procedures and using appropriate tools, businesses can improve the accuracy, consistency, and reliability of their data, which is essential for effective decision-making and operations.

Are there existing procedures for data cleaning and validation?

The frequency of data quality assessment and maintenance varies depending on the specific needs of the business, the type of data, and its usage. However, it is generally important to assess and maintain data quality regularly to ensure the accuracy, reliability, and relevance of the data. Here are some common practices and recommended frequencies for data quality assessment and maintenance:

Ongoing or Real-Time Monitoring

1. **Continuous Monitoring:**
   - For critical data systems, continuous monitoring tools can be used to track data quality in real-time and alert administrators to issues as they arise.
   - Example: Monitoring transaction data in financial systems to detect anomalies or errors immediately.
2. **Automated Validation:**
   - Implementing automated validation checks that run whenever data is entered, updated, or transferred between systems.
   - Example: Validating data input in real-time during customer registration or order processing.

Daily Assessments

1. **Daily Data Quality Checks:**
   - Regularly scheduled scripts or processes that check for common data quality issues daily.
   - Example: Daily batch jobs that clean and validate sales data before it is loaded into the data warehouse.

Weekly Assessments

1. **Weekly Data Profiling:**
   - Conducting a weekly review of data to identify trends, patterns, and anomalies that may indicate data quality issues.
   - Example: Weekly profiling of customer data to identify incomplete or inconsistent records.

Monthly Assessments

1. **Monthly Data Audits:**
    - Comprehensive audits that evaluate the overall quality of data and compliance with data governance policies.
    - Example: Monthly audits of financial data to ensure accuracy and compliance with accounting standards.

Quarterly Assessments

- **Quarterly Data Quality Reviews:**
    - Formal reviews of data quality metrics and key performance indicators (KPIs) conducted quarterly.
    - Example: Quarterly meetings to review data quality reports and address systemic issues affecting data integrity.
- **Data Cleansing Campaigns:**
    - Intensive data cleansing activities focused on specific datasets or systems.
    - Example: Quarterly campaigns to clean and update contact information in the customer relationship management (CRM) system

Annual Assessments

1. **Annual Data Quality Assessments:**
    - Comprehensive evaluation of the organization's data quality practices, policies, and overall data health.
    - Example: Annual assessments that benchmark data quality against industry standards and best practices.
2. **Data Governance Reviews:**
    - Annual reviews of data governance frameworks, policies, and procedures to ensure they remain effective and aligned with business objectives.
    - Example: Annual governance reviews that involve stakeholders from various departments to ensure data quality initiatives are aligned with organizational goals.

Event-Driven Assessments

1. **Post-Migration Checks:**
    - Assessing data quality after major data migrations, system upgrades, or integrations.
    - Example: Data validation following the migration of data to a new enterprise resource planning (ERP) system.
2. **Regulatory Compliance Audits:**
    - Conducting data quality assessments in response to regulatory requirements or audits.
    - Example: Ensuring data quality and compliance before a regulatory audit in the healthcare or finance industry.

Best Practices for Maintaining Data Quality

1. **Implement Data Quality Metrics and KPIs:**
   - Establishing clear metrics and KPIs to measure data quality and track improvements over time.
   - Example: Tracking metrics such as data accuracy, completeness, consistency, and timeliness.
2. **Utilize Data Quality Tools:**
   - Leveraging data quality and data governance tools to automate and streamline data quality assessments.
   - Example: Using tools like Talend, Informatica, or Collibra for automated data quality management.
3. **Engage Stakeholders:**
   - Involving key stakeholders in data quality initiatives to ensure buy-in and accountability.
   - Example: Forming a data governance committee with representatives from different departments.

By adopting a structured approach to data quality assessment and maintenance, businesses can ensure that their data remains accurate, reliable, and valuable for decision-making and operations.

How frequently is data quality assessed and maintained?

Data Infrastructure: Data Accessibility

Access to data within a business is typically controlled and managed based on roles and responsibilities, ensuring that the right people have access to the appropriate data while maintaining security and compliance. Here are the key roles and types of access typically involved in a business:

Key Roles and Types of Access

1. **Executives and Senior Management:**
   - **Access Level:** High-level access to summary and aggregate data across the organization.
   - **Purpose:** Strategic decision-making, performance monitoring, and business planning.
   - **Examples:** CEO, CFO, COO.
2. **Department Heads and Managers:**
   - **Access Level:** Access to detailed data relevant to their specific departments or teams.
   - **Purpose:** Operational management, performance tracking, and departmental planning.
   - **Examples:** Sales Manager, Marketing Director, HR Manager.
3. **Data Analysts and Data Scientists:**
   - **Access Level:** Broad access to various datasets for analysis, modeling, and reporting.
   - **Purpose:** Data analysis, business intelligence, predictive modeling, and data-driven insights.
   - **Examples:** Business Analyst, Data Scientist, BI Developer.
4. **IT and Data Management Teams:**

- **Access Level:** Full access to data systems and infrastructure for maintenance, security, and administration.
- **Purpose:** Data management, system administration, security enforcement, and technical support.
- **Examples:** Database Administrator, IT Manager, Data Engineer.

5. **Operational Staff:**
   - **Access Level:** Access to specific data needed for day-to-day tasks and operations.
   - **Purpose:** Executing business processes, customer service, transaction processing.
   - **Examples:** Customer Service Representative, Sales Associate, Operations Coordinator.
6. **Compliance and Audit Teams:**
   - **Access Level:** Access to data necessary for compliance checks, audits, and regulatory reporting.
   - **Purpose:** Ensuring compliance with legal, regulatory, and internal policies.
   - **Examples:** Compliance Officer, Internal Auditor, Risk Manager.
7. **External Partners and Vendors:**
   - **Access Level:** Restricted access to specific data required to fulfill contractual obligations or provide services.
   - **Purpose:** Collaboration, service provision, and third-party integrations.
   - **Examples:** Cloud Service Providers, Marketing Agencies, Outsourced IT Support.

Access Control Mechanisms

1. **Role-Based Access Control (RBAC):**
   - Assigns access permissions based on user roles within the organization, ensuring that employees only access data necessary for their job functions.
2. **Data Classification and Sensitivity Levels:**
   - Classifies data based on its sensitivity and criticality (e.g., public, internal, confidential, restricted) and applies appropriate access controls.
3. **Least Privilege Principle:**
   - Grants users the minimum level of access necessary to perform their duties, reducing the risk of unauthorized data access.
4. **Access Audits and Monitoring:**
   - Regularly reviews and monitors access logs to detect and respond to unauthorized access or unusual activities.
5. **Data Encryption and Security Measures:**
   - Implements encryption and other security measures to protect sensitive data both at rest and in transit.
6. **User Authentication and Authorization:**
   - Uses strong authentication methods (e.g., multi-factor authentication) and authorization processes to verify user identities and grant access.
7. **Data Governance Policies:**
   - Establishes clear policies and procedures for data access, handling, and sharing, ensuring compliance with regulatory requirements and internal standards.

By implementing these access control mechanisms and defining clear roles and responsibilities, businesses can ensure that data is accessible to those who need it while protecting it from unauthorized access and potential breaches.

Who has access to the data within your organization?

Businesses may face several restrictions and challenges in accessing necessary data, which can impact their operations, decision-making, and overall efficiency. Here are some common restrictions and challenges:

Legal and Regulatory Restrictions

1. **Data Privacy Laws:**
    - Compliance with laws such as GDPR, CCPA, and HIPAA can restrict access to personal and sensitive data. Businesses must ensure that data access complies with these regulations to avoid legal penalties.
2. **Industry-Specific Regulations:**
    - Certain industries have specific regulations that dictate how data must be handled and accessed, such as financial regulations (e.g., SOX) or healthcare regulations (e.g., HITECH Act).

Security and Privacy Concerns

1. **Data Security:**
    - Ensuring data security can restrict access, as businesses must implement strong authentication, encryption, and access control measures to protect data from breaches and unauthorized access.
2. **Internal Security Policies:**
    - Organizations may have internal policies that restrict data access to maintain confidentiality and integrity, limiting who can access certain types of data.

Technical Challenges

1. **Data Silos:**
    - Data stored in separate systems or departments can create silos, making it difficult to access and integrate data across the organization.
2. **Legacy Systems:**
    - Older systems may not support modern data access methods, making it challenging to retrieve and use data efficiently.

3. **Data Integration:**
    - Integrating data from multiple sources with different formats, structures, and standards can be complex and time-consuming.

Data Quality Issues

1. **Incomplete or Inaccurate Data:**
    - Poor data quality, including incomplete, inaccurate, or inconsistent data, can hinder effective access and analysis.
2. **Data Duplication:**
    - Duplicate records can create confusion and complicate data access processes.

Organizational Challenges

1. **Lack of Data Governance:**
    - Without proper data governance frameworks, businesses may struggle to manage and access data effectively.
2. **Resource Constraints:**
    - Limited resources, such as budget, staff, or technology, can restrict a business's ability to access and manage data.
3. **Change Management:**
    - Resistance to change and lack of training can hinder the adoption of new data access technologies and processes.

Performance and Scalability Issues

1. **System Performance:**
    - High data volumes and complex queries can impact system performance, leading to slow access times and delays.
2. **Scalability:**
    - Ensuring that data access solutions can scale with the growing amount of data and user demands can be challenging.

External Dependencies

1. **Third-Party Vendors:**
    - Relying on third-party vendors for data storage or management can create dependency issues, especially if the vendor experiences outages or other problems.
2. **Data Sharing Agreements:**
    - Legal agreements with partners or suppliers may limit how data can be accessed and used.

Data Accessibility

1. **User Accessibility:**

    - Ensuring that data is accessible to all relevant users while maintaining security can be challenging, particularly in large organizations.
2. **Technical Skills:**
    - Users may lack the technical skills to effectively access and use data, necessitating additional training and support.

Best Practices to Mitigate Challenges

1. **Implement Robust Data Governance:**
    - Establish clear policies and procedures for data access, management, and usage.
2. **Enhance Data Security Measures:**
    - Use strong encryption, authentication, and access control mechanisms to protect data.
3. **Invest in Modern Data Infrastructure:**
    - Upgrade legacy systems and invest in scalable, integrated data management solutions.
4. **Ensure Compliance:**
    - Stay informed about relevant laws and regulations and ensure compliance through regular audits and updates.
5. **Improve Data Quality:**
    - Implement data cleaning, validation, and quality assurance processes to maintain high data quality.
6. **Promote Collaboration:**
    - Break down data silos and encourage collaboration across departments to improve data access and integration.
7. **Provide Training and Support:**
    - Offer training programs to enhance data literacy and technical skills among employees.

By addressing these challenges and implementing best practices, businesses can improve their ability to access and utilize necessary data effectively.

Are there any restrictions or challenges in accessing your necessary data?

Data access in a business is monitored and controlled through a combination of policies, technologies, and practices designed to ensure that only authorized individuals can access specific data. This is crucial for maintaining data security, privacy, and compliance with legal and regulatory requirements. Here are the key methods used to monitor and control data access:

Access Control Mechanisms

1. **Role-Based Access Control (RBAC):**
   - Access permissions are assigned based on the user's role within the organization. Each role has a predefined set of access rights, ensuring that employees can access only the data necessary for their job functions.
2. **Attribute-Based Access Control (ABAC):**
   - Access decisions are based on attributes (user, resource, environment) rather than roles alone. This allows for more granular access control policies.
3. **Discretionary Access Control (DAC):**
   - Data owners determine who can access their data and what permissions they have. This model provides flexibility but requires careful management to avoid unauthorized access.
4. **Mandatory Access Control (MAC):**
   - Access rights are assigned based on regulations or policies, often used in environments requiring high security, such as government or military institutions.

Authentication and Authorization

1. **Multi-Factor Authentication (MFA):**
   - Users must provide multiple forms of verification (e.g., password, biometric, security token) to access data, adding an extra layer of security.
2. **Single Sign-On (SSO):**
   - Users can access multiple systems with one set of credentials, simplifying access management and improving security.
3. **Access Requests and Approvals:**
   - A formal process for requesting and approving data access ensures that only authorized users gain access to sensitive information.

Monitoring and Auditing

1. **Access Logs:**
   - Detailed logs record who accessed data, when, and what actions were taken. These logs are essential for tracking user activity and identifying unauthorized access attempts.
2. **Audit Trails:**
   - Comprehensive records of all data access and changes, used to ensure compliance with policies and regulations and to investigate incidents.
3. **Real-Time Monitoring:**
   - Continuous monitoring of data access to detect and respond to suspicious activities or potential security breaches immediately.

4. **Data Loss Prevention (DLP):**
   - Tools and policies designed to prevent unauthorized data transfers or leaks, often through monitoring and controlling data flow.

Data Governance and Policies

1. **Data Classification:**
   - Data is classified based on its sensitivity and criticality (e.g., public, internal, confidential, restricted), with access controls tailored to each classification level.
2. **Data Access Policies:**
   - Clear policies define who can access what data and under what conditions. These policies are enforced through technical controls and regularly reviewed and updated.
3. **Least Privilege Principle:**
   - Users are granted the minimum level of access necessary to perform their job functions, reducing the risk of unauthorized access.
4. **Periodic Access Reviews:**
   - Regular reviews of access permissions to ensure they are still appropriate and necessary, removing access for users who no longer need it.

Technical Controls

1. **Encryption:**
   - Data is encrypted both at rest and in transit to protect it from unauthorized access.
2. **Intrusion Detection and Prevention Systems (IDPS):**
   - Systems that monitor network and system activities for malicious actions or policy violations, capable of taking automatic actions to prevent breaches.
3. **Endpoint Security:**
   - Measures to secure endpoints (e.g., computers, mobile devices) that access data, including antivirus, firewalls, and device management tools.

Compliance and Training

1. **Regulatory Compliance:**
   - Ensuring data access controls comply with relevant laws and regulations, such as GDPR, HIPAA, and CCPA.
2. **Employee Training:**
   - Regular training programs to educate employees about data access policies, security best practices, and the importance of protecting sensitive information.

Incident Response

1. **Incident Detection and Response:**
   - Procedures for detecting, investigating, and responding to data access incidents, including breach notification and remediation steps.

By implementing these methods, businesses can effectively monitor and control data access, ensuring that data is protected from unauthorized access while remaining accessible to those who need it for their work.

How is data access monitored and controlled?

Data Infrastructure: Data Integration

Determining how well your data sources are integrated within your business involves evaluating several key aspects of data integration, including data consistency, accessibility, usability, and overall system performance. Here are some methods and metrics to assess the quality of your data integration:

1. Data Consistency and Accuracy

- **Data Quality Reports:** Regularly generate and review reports that highlight data quality issues such as duplicates, inconsistencies, and inaccuracies across integrated systems.

- **Data Validation Rules:** Implement and monitor data validation rules to ensure that data conforms to expected formats and values as it moves between systems.
- **Error Rates:** Track the frequency and types of errors that occur during data integration processes.

2. Data Accessibility and Timeliness

- **Data Availability:** Measure the availability and accessibility of integrated data sources to end-users and applications. High availability indicates better integration.
- **Latency:** Evaluate the time it takes for data to move between systems and become available for use. Lower latency suggests more efficient integration.
- **Real-Time Integration:** Assess the capability to access and utilize data in real-time or near real-time, which is critical for certain business functions.

3. Data Usability and User Satisfaction

- **User Feedback:** Collect feedback from end-users on the usability and reliability of the integrated data. User satisfaction surveys can provide insights into perceived integration effectiveness.
- **User Adoption Rates:** Monitor the adoption rates of integrated data systems and tools. High adoption indicates that users find the integration beneficial and reliable.
- **Ease of Use:** Evaluate how easily users can access and work with integrated data, including the intuitiveness of interfaces and availability of training resources.

4. System Performance and Reliability

- **System Downtime:** Track the frequency and duration of system outages that affect data integration. Lower downtime indicates more reliable integration.
- **Performance Metrics:** Monitor system performance metrics such as processing speed, query response times, and load times for integrated data systems.
- **Scalability:** Assess the ability of your data integration solutions to scale with increasing data volumes and user demands without degradation in performance.

5. Integration Coverage and Completeness

- **Data Mapping:** Review the completeness and accuracy of data mappings between different sources and systems. Well-documented and comprehensive mappings indicate better integration.
- **Coverage Analysis:** Ensure that all relevant data sources are integrated and that there are no significant gaps in the data being integrated.
- **Redundancy and Duplication:** Evaluate the level of redundancy and duplication in integrated data. Effective integration should minimize unnecessary redundancy.

6. Compliance and Security

- **Compliance Audits:** Conduct regular audits to ensure that data integration processes comply with relevant legal, regulatory, and internal policies.
- **Security Measures:** Evaluate the effectiveness of security measures in place to protect data during integration, including encryption, access controls, and monitoring.

7. Key Performance Indicators (KPIs)

- **Integration KPIs:** Define and track KPIs specific to data integration, such as the number of successful data transfers, integration error rates, and data synchronization frequency.
- **Business Impact KPIs:** Measure the impact of data integration on key business outcomes, such as decision-making speed, operational efficiency, and customer satisfaction.

Tools and Technologies for Assessing Data Integration

- **Data Integration Platforms:** Use data integration platforms and tools (e.g., Informatica, Talend, Microsoft SSIS) that offer built-in monitoring and reporting features.
- **Data Quality Tools:** Employ data quality tools to continuously assess and improve data quality across integrated sources.
- **BI and Analytics Tools:** Leverage business intelligence (BI) and analytics tools to visualize and analyze the effectiveness of data integration.

By systematically evaluating these aspects, businesses can gain a comprehensive understanding of how well their data sources are integrated and identify areas for improvement to enhance data quality, accessibility, and overall business value.

How well are your data sources integrated?

Identifying data silos and disconnected data sources in your business involves several steps, including analyzing data flow, assessing data accessibility, and gathering feedback from stakeholders. Here's a structured approach to determine if data silos exist in your organization:

1. Conduct a Data Inventory

- **Catalog Data Sources:** Create a comprehensive inventory of all data sources within the organization, including databases, applications, spreadsheets, and external data feeds.
- **Metadata Analysis:** Collect metadata for each data source, including data ownership, data types, volume, and frequency of updates.

2. Map Data Flows

- **Data Flow Diagrams:** Develop data flow diagrams to visualize how data moves between different systems and departments. Identify any points where data does not flow freely between systems.
- **Integration Points:** Document where and how data integration occurs, noting any systems that are not integrated.

3. Assess Data Accessibility

- **User Access Surveys:** Survey employees to understand which data sources they can access and which they cannot. Identify any data sources that are not widely accessible or are restricted to specific departments.
- **Access Logs:** Analyze access logs and permissions settings to see who has access to which data. Look for patterns indicating that certain data is siloed within specific teams or roles.

4. Evaluate Data Usage and Redundancy

- **Usage Patterns:** Review how data is used across the organization. Identify any duplication of data efforts, such as multiple departments maintaining separate copies of the same data.
- **Redundant Systems:** Check for redundant systems or processes that suggest data is being maintained separately in silos.

5. Gather Stakeholder Feedback

- **Interviews and Workshops:** Conduct interviews and workshops with key stakeholders from different departments to understand their data needs and challenges. Ask specifically about any difficulties in accessing or sharing data.
- **Feedback on Pain Points:** Collect feedback on pain points related to data access and integration. Look for common themes indicating data silos.

6. Analyze Data Consistency and Quality

- **Data Quality Reports:** Generate reports on data quality across different systems. Inconsistencies in data quality can indicate siloed data management.
- **Data Reconciliation:** Attempt to reconcile similar data sets from different sources. Difficulty in reconciling data can reveal silos.

7. Technology and Integration Assessment

- **Integration Tools:** Assess the tools and technologies used for data integration. Identify any gaps in integration capabilities or technologies that are not being utilized.
- **System Compatibility:** Evaluate the compatibility of different systems and platforms. Incompatibilities can lead to data silos.

8. Review Data Governance Practices

- **Governance Policies:** Review data governance policies to ensure they promote data sharing and integration across the organization.
- **Data Stewardship:** Identify data stewards and assess their roles in managing and sharing data. Lack of centralized data stewardship can contribute to silos.

9. Perform a Gap Analysis

- **Identify Gaps:** Conduct a gap analysis to compare the current state of data integration with the desired state. Highlight areas where data silos exist and where integration is lacking.
- **Prioritize Issues:** Prioritize the identified issues based on their impact on business operations and decision-making.

10. Implement Continuous Monitoring

- **Monitoring Tools:** Deploy monitoring tools to continuously track data access, data flows, and integration points. Regularly review the monitoring results to detect new or persistent silos.
- **KPIs and Metrics:** Establish key performance indicators (KPIs) and metrics to measure data integration and accessibility. Regularly review these metrics to ensure continuous improvement.

11. Tools and Techniques

- **Data Integration Platforms:** Utilize data integration platforms like Talend, Informatica, or Microsoft SSIS to facilitate and monitor data integration efforts.
- **Business Intelligence Tools:** Leverage BI tools like Tableau, Power BI, or Looker to create visualizations that highlight data access and usage patterns.
- **Data Governance Tools:** Implement data governance tools to manage and enforce data sharing policies.

By following these steps, businesses can systematically identify data silos and disconnected data sources, allowing them to take corrective actions to improve data integration and accessibility.

Are there any silos or disconnected data sources?

Data integration is a critical aspect of managing data effectively in a business. Various tools and platforms can help integrate data from multiple sources, ensuring it is accessible, consistent, and useful for analysis and decision-making. Here are some of the widely used tools and platforms for data integration:

1. ETL (Extract, Transform, Load) Tools

- **Informatica PowerCenter:** A comprehensive data integration tool that offers high-performance data integration, data quality, and data governance capabilities.
- **Talend:** An open-source data integration platform that provides ETL, data profiling, and big data integration functionalities.
- **Microsoft SQL Server Integration Services (SSIS):** A platform for building enterprise-level data integration and data transformation solutions.

2. Data Integration Platforms

- **Apache Nifi:** A powerful data integration tool that supports data routing, transformation, and system mediation logic.
- **MuleSoft Anypoint Platform:** An integration platform that provides API-led connectivity to enable applications, data, and devices to be connected seamlessly.
- **Dell Boomi:** A cloud-based integration platform that supports connecting applications and data across environments with low-code development.

3. Data Warehousing Solutions

- **Amazon Redshift:** A fully managed data warehouse service that makes it simple and cost-effective to analyze all your data using standard SQL and existing BI tools.
- **Google BigQuery:** A serverless, highly scalable, and cost-effective multi-cloud data warehouse designed for business agility.
- **Snowflake:** A cloud data platform that provides data warehousing, data lakes, and data sharing capabilities.

4. iPaaS (Integration Platform as a Service)

- **Zapier:** An easy-to-use tool for integrating apps and automating workflows, suitable for small to medium-sized businesses.
- **Workato:** An enterprise automation platform that integrates applications and automates workflows.
- **Tray.io:** A flexible iPaaS platform for integrating applications and automating complex workflows.

5. Cloud Data Integration

- **Azure Data Factory:** A cloud-based data integration service that allows you to create data-driven workflows for orchestrating data movement and data transformation.
- **Google Cloud Dataflow:** A fully managed service for stream and batch processing of data.
- **AWS Glue:** A serverless data integration service that makes it easy to discover, prepare, and combine data for analytics, machine learning, and application development.

6. API Management and Integration

- **Postman:** A popular tool for API development and testing, which can also be used to integrate APIs for data exchange between systems.
- **Apigee:** A full lifecycle API management platform that enables API integration, monitoring, and security.

7. Data Virtualization

- **Denodo:** A data virtualization platform that provides real-time access to data across multiple sources without the need for physical data movement.
- **TIBCO Data Virtualization:** A tool that allows you to create a unified, logical view of data from multiple sources.

8. Business Intelligence and Analytics Platforms

- **Tableau:** A powerful BI tool that can integrate data from various sources to provide interactive, visual analytics.
- **Power BI:** A business analytics tool by Microsoft that provides data visualization and integration with various data sources.
- **Looker:** A data platform that offers data integration and analytics to explore and share insights across your organization.

9. Data Quality and Governance Tools

- **Collibra:** A data governance platform that helps manage data assets, ensuring data quality and compliance.

- **Alation:** A data catalog and governance tool that supports data discovery, quality, and compliance.

10. Custom Solutions and Scripting

- **Python with Pandas and SQLAlchemy:** For custom data integration tasks, Python libraries like Pandas and SQLAlchemy can be used to extract, transform, and load data programmatically.
- **Apache Spark:** An open-source unified analytics engine for large-scale data processing, suitable for big data integration tasks.

By selecting the right combination of these tools and platforms, businesses can effectively integrate data from various sources, ensuring it is clean, consistent, and ready for analysis and decision-making.

What tools or platforms do you use for data integration?

Technology Stack: Hardware Infrastructure

The types of hardware used in a business entity can vary significantly depending on the size of the business, its industry, and its specific needs. However, most businesses typically use a combination of the following hardware components to support their operations:

Servers

1. **Application Servers:**
    - Host applications and software used by the business, enabling multiple users to access and run these applications simultaneously.
    - Examples: Web servers, email servers, ERP servers.
2. **Database Servers:**
    - Store and manage databases, allowing for efficient data retrieval, storage, and management.
    - Examples: SQL servers, NoSQL servers.
3. **File Servers:**
    - Provide a centralized location for storing and sharing files across the network.
    - Examples: Network Attached Storage (NAS), dedicated file servers.
4. **Mail Servers:**
    - Manage and store email communications for the business.
    - Examples: Microsoft Exchange Server, Google Workspace.
5. **Web Servers:**
    - Host websites and web applications, making them accessible over the internet.
    - Examples: Apache HTTP Server, Nginx.

Storage Devices

1. **Direct Attached Storage (DAS):**
    - Storage devices that are directly connected to a server or workstation, such as internal hard drives and external USB drives.
2. **Network Attached Storage (NAS):**
    - Dedicated storage devices connected to the network, providing centralized storage and file-sharing capabilities.
    - Examples: Synology NAS, QNAP NAS.
3. **Storage Area Network (SAN):**
    - High-speed network that connects multiple storage devices to servers, allowing for large-scale storage management and access.
    - Examples: Dell EMC PowerMax, HPE 3PAR.
4. **Cloud Storage:**
    - Storage services provided by third-party cloud providers, offering scalable and flexible storage solutions.
    - Examples: Amazon S3, Microsoft Azure Blob Storage, Google Cloud Storage.

Networking Hardware

1. **Routers:**
    - Direct data traffic between different networks, connecting the business network to the internet or other networks.
    - Examples: Cisco routers, Juniper routers.
2. **Switches:**

   - Connect multiple devices within a local area network (LAN), enabling communication between them.
   - Examples: Cisco Catalyst switches, HP Aruba switches.
3. **Firewalls:**
   - Protect the business network from unauthorized access and cyber threats by monitoring and controlling incoming and outgoing network traffic.
   - Examples: Palo Alto Networks firewalls, Fortinet FortiGate.
4. **Access Points:**
   - Provide wireless connectivity to the network, enabling devices to connect via Wi-Fi.
   - Examples: Ubiquiti UniFi access points, Cisco Meraki access points.

Computing Devices

1. **Desktop Computers:**
   - Used by employees for day-to-day tasks, such as office work, software development, and design.
   - Examples: Dell OptiPlex, HP EliteDesk.
2. **Laptops:**
   - Portable computing devices for employees who need mobility.
   - Examples: Lenovo ThinkPad, Apple MacBook Pro.
3. **Workstations:**
   - High-performance computers designed for intensive tasks like graphic design, video editing, and engineering simulations.
   - Examples: HP Z Series, Dell Precision.

Peripheral Devices

1. **Printers and Scanners:**
   - Used for printing documents and scanning physical documents into digital format.
   - Examples: HP LaserJet, Canon imageRUNNER.
2. **Monitors:**
   - Display screens used with desktop computers and workstations.
   - Examples: Dell UltraSharp, Samsung monitors.
3. **Keyboards and Mice:**
   - Input devices used with desktop computers and workstations.

Backup and Recovery Hardware

1. **Backup Servers:**
   - Dedicated servers for managing data backup processes and storing backup copies.
   - Examples: Veeam Backup & Replication, Veritas Backup Exec.
2. **Tape Drives:**
   - Used for long-term data storage and archival purposes.
   - Examples: IBM TS4500, HPE StoreEver.
3. **External Hard Drives:**

- Portable storage devices for backup and data transfer.
- Examples: Western Digital My Passport, Seagate Backup Plus.

Specialized Hardware

1. **Point of Sale (POS) Systems:**
   - Used in retail and hospitality businesses for processing sales transactions.
   - Examples: Square POS, Clover POS.
2. **Security Cameras and DVRs/NVRs:**
   - Used for physical security monitoring and recording.
   - Examples: Hikvision cameras, Dahua NVRs.

Virtualization and Cloud Hardware

1. **Virtualization Servers:**
   - Servers that run virtualization software, allowing multiple virtual machines (VMs) to run on a single physical server.
   - Examples: VMware ESXi hosts, Microsoft Hyper-V servers.
2. **Cloud Gateway Appliances:**
   - Devices that facilitate connectivity between on-premises infrastructure and cloud services.
   - Examples: AWS Storage Gateway, Azure Stack.

By leveraging these hardware components, businesses can create a robust IT infrastructure that supports their operational needs, enhances productivity, and ensures data security and accessibility.

What types of hardware are you currently using?

Are there any plans for upgrading or expanding your hardware infrastructure?

Determining if your current hardware setup can support AI applications involves evaluating several key aspects of your infrastructure, including computing power, storage, networking capabilities, and compatibility with AI frameworks. Here are some steps to help you assess your hardware readiness for AI applications:

1. Evaluate Computing Power

CPUs and GPUs:

- **CPU Capability:** Ensure that your CPUs have enough cores and processing power to handle AI workloads. AI applications often require high-performance, multi-core processors.
- **GPU Availability:** AI applications, especially those involving deep learning, benefit significantly from GPUs due to their parallel processing capabilities. Check if you have powerful GPUs (e.g., NVIDIA Tesla, Quadro, or AMD Radeon Instinct) in your setup.
- **TPUs and FPGAs:** Specialized hardware like Tensor Processing Units (TPUs) and Field-Programmable Gate Arrays (FPGAs) can also accelerate AI tasks. Assess if these are needed and available.

Memory (RAM):

- **Sufficient RAM:** AI models, particularly large neural networks, require substantial amounts of RAM. Ensure your servers or workstations have enough memory to load and process large datasets and models.

2. Assess Storage and Data Handling

Storage Capacity:

- **Data Storage:** Ensure you have adequate storage capacity for large datasets. AI applications often require extensive datasets for training, which can consume significant storage space.
- **High-Speed Storage:** Fast storage solutions such as SSDs or NVMe drives can improve data read/write speeds, which is critical for AI training and inference tasks.

Data Throughput:

- **I/O Performance:** Evaluate the input/output performance of your storage systems to ensure they can handle the high data throughput required by AI applications.

3. Network Capabilities

Bandwidth:

- **Network Bandwidth:** High bandwidth is essential for transferring large datasets between storage, compute nodes, and other network components. Ensure your network infrastructure can support high-speed data transfers.

Latency:

- **Low Latency:** Low network latency is crucial for distributed AI training and real-time inference applications. Assess your network's latency and optimize if necessary.

4. Compatibility with AI Frameworks

Software and Frameworks:

- **AI Framework Support:** Ensure your hardware is compatible with popular AI frameworks such as TensorFlow, PyTorch, Keras, and others. This includes compatibility with necessary libraries and drivers.
- **CUDA and cuDNN Support:** For NVIDIA GPUs, ensure that CUDA and cuDNN are supported and properly configured, as they are essential for accelerating AI computations.

5. Scalability and Flexibility

Scalability:

- **Expandability:** Assess whether your current hardware setup can be easily scaled up or out to accommodate growing AI workloads. This includes adding more GPUs, increasing storage, or enhancing network infrastructure.

Flexibility:

- **Hybrid and Multi-Cloud Readiness:** Evaluate if your setup can leverage cloud resources for AI workloads, providing flexibility and additional compute power as needed.

6. Performance Benchmarks and Testing

Benchmarking:

- **Run Benchmarks:** Use AI-specific benchmarking tools to evaluate the performance of your current hardware setup. Tools like MLPerf can provide insights into how well your hardware handles AI tasks.
- **Stress Testing:** Perform stress tests to understand the limits of your hardware under peak AI workloads.

7. Specialized Hardware Considerations

AI-Specific Hardware:

- **AI Accelerators:** Evaluate the need for AI-specific accelerators like Google's TPUs or Intel's Nervana processors, which can significantly speed up AI tasks.
- **Edge Devices:** If deploying AI at the edge, ensure edge devices have the necessary computational power and connectivity to support AI inference.

Steps to Assess Your Current Hardware Setup

1. **Inventory Your Hardware:** Catalog your existing hardware, including CPUs, GPUs, storage devices, and network equipment.
2. **Analyze Workload Requirements:** Understand the specific requirements of your AI applications, including computational needs, data storage, and processing capabilities.
3. **Run Diagnostic Tools:** Use tools like NVIDIA's nvidia-smi for GPU diagnostics, CPU benchmarking tools, and network performance analyzers.
4. **Consult Documentation:** Review documentation for your AI frameworks to ensure compatibility and optimal configuration.
5. **Engage with Experts:** Consult with AI and IT infrastructure experts to get tailored recommendations based on your specific use case and current setup.

By following these steps, you can determine if your current hardware setup is capable of supporting AI applications or if upgrades and optimizations are necessary to meet your AI goals.

How scalable is your current hardware setup to support AI applications?

Technology Stack: Software Infrastructure

A business typically uses a variety of software platforms and applications to support its operations, enhance productivity, and manage different aspects of the business. Here are some common software platforms and applications that a business might have in place:

Business Management

1. **Enterprise Resource Planning (ERP):**
   - **Examples:** QuickBooks Online, Zoho Books, Xero
   - **Purpose:** Manage accounting, invoicing, inventory, and financial reporting.
2. **Customer Relationship Management (CRM):**
   - **Examples:** Salesforce, HubSpot CRM, Zoho CRM
   - **Purpose:** Manage customer interactions, sales pipelines, and marketing campaigns.

Productivity and Collaboration

3. **Office Suites:**
   - **Examples:** Microsoft 365, Google Workspace
   - **Purpose:** Provide tools for word processing, spreadsheets, presentations, and email.
4. **Project Management:**
   - **Examples:** Trello, Asana, Monday.com
   - **Purpose:** Plan, track, and collaborate on projects and tasks.
5. **Communication Tools:**
   - **Examples:** Slack, Microsoft Teams, Zoom
   - **Purpose:** Facilitate team communication, video conferencing, and instant messaging.

Marketing and Sales

6. **Email Marketing:**
   - **Examples:** Mailchimp, Constant Contact, Sendinblue
   - **Purpose:** Create, send, and track email marketing campaigns.
7. **Social Media Management:**
   - **Examples:** Hootsuite, Buffer, Sprout Social
   - **Purpose:** Schedule, manage, and analyze social media posts and engagements.
8. **E-commerce Platforms:**
   - **Examples:** Shopify, WooCommerce, BigCommerce
   - **Purpose:** Manage online stores, product listings, orders, and payments.

Website and Content Management

9. **Content Management Systems (CMS):**
   - **Examples:** WordPress, Wix, Squarespace
   - **Purpose:** Create and manage website content, blogs, and landing pages.
10. **Analytics and SEO Tools:**
   - **Examples:** Google Analytics, SEMrush, Moz
   - **Purpose:** Track website performance, analyze traffic, and optimize for search engines.

Financial and Operational Tools

11. **Accounting Software:**
   - **Examples:** QuickBooks, FreshBooks, Wave
   - **Purpose:** Manage invoicing, expenses, payroll, and financial reporting.
12. **Payment Processing:**
   - **Examples:** PayPal, Square, Stripe
   - **Purpose:** Process payments, manage transactions, and handle point-of-sale (POS) operations.
13. **Inventory Management:**
   - **Examples:** TradeGecko, Cin7, Zoho Inventory
   - **Purpose:** Track inventory levels, manage orders, and streamline supply chain operations.

Security and Backup

14. **Cybersecurity:**
   - **Examples:** Norton Small Business, McAfee Total Protection, Bitdefender
   - **Purpose:** Protect against malware, phishing, and other cybersecurity threats.
15. **Backup and Recovery:**
   - **Examples:** Carbonite, Backblaze, Acronis
   - **Purpose:** Backup important data and ensure data recovery in case of hardware failure or data loss.

Human Resources and Payroll

16. **HR Management:**
    - **Examples:** Gusto, BambooHR, Zenefits
    - **Purpose:** Manage employee records, onboarding, benefits, and payroll.
17. **Time Tracking:**
    - **Examples:** Toggl, Harvest, TSheets
    - **Purpose:** Track employee hours, project time, and productivity.

Customer Support

18. **Help Desk Software:**
    - **Examples:** Zendesk, Freshdesk, Help Scout
    - **Purpose:** Manage customer support tickets, live chat, and service requests.

Specialized Industry Tools

19. **Point of Sale (POS) Systems:**
    - **Examples:** Square POS, Lightspeed, Clover
    - **Purpose:** Handle in-store sales transactions, inventory management, and customer data.
20. **Booking and Scheduling:**
    - **Examples:** Calendly, Acuity Scheduling, Booksy
    - **Purpose:** Manage appointments, bookings, and client schedules.

Other Utilities

21. **Document Management:**
    - **Examples:** Dropbox, Google Drive, Microsoft OneDrive
    - **Purpose:** Store, share, and manage documents and files.
22. **Graphic Design:**
    - **Examples:** Adobe Creative Cloud, Canva, Affinity Designer
    - **Purpose:** Create and edit graphics, marketing materials, and visual content.

By leveraging these software platforms and applications, businesses can streamline their operations, improve efficiency, and enhance their overall productivity and customer service.

What software platforms and applications are currently in place?

Are there any AI-specific software tools or frameworks in place?

How frequently are software systems updated or upgraded?

Technology Stack: Network Infrastructure

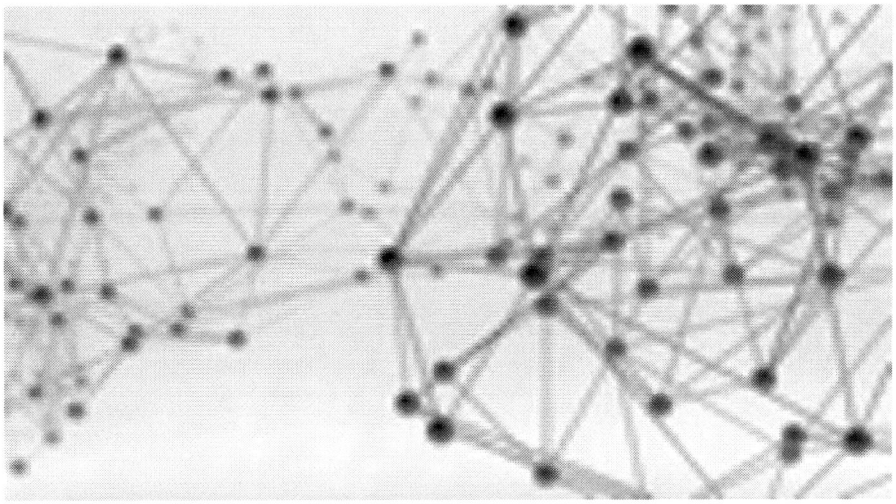

Determining the robustness and reliability of your network infrastructure involves assessing various components, performance metrics, and security measures. Here are steps and methods to evaluate the network infrastructure in a business:

1. Performance Metrics

Network Speed and Throughput:

- **Bandwidth Tests:** Use tools like Speedtest by Ookla, iPerf, or NetFlow to measure network bandwidth and throughput.

- **Latency and Jitter:** Assess the time it takes for data packets to travel between devices. High latency and jitter can indicate network issues.

Network Availability and Uptime:

- **Uptime Monitoring:** Tools like Pingdom, Nagios, or SolarWinds can track network uptime and alert you to outages.
- **Service Level Agreements (SLAs):** Check if the actual network uptime meets the SLAs promised by your service providers.

2. Network Configuration and Design

Network Topology:

- **Documentation Review:** Ensure that network diagrams and documentation are up to date, reflecting the current network layout.
- **Redundancy:** Verify that your network has redundant paths and failover mechanisms to maintain connectivity during failures.

Hardware and Devices:

- **Equipment Quality:** Assess the quality and age of network hardware (routers, switches, firewalls). Outdated or low-quality hardware can be a weak link.
- **Firmware Updates:** Ensure all devices have the latest firmware updates to avoid security vulnerabilities and performance issues.

3. Security Measures

Firewalls and Intrusion Detection:

- **Firewall Configuration:** Review firewall rules and configurations to ensure they are optimized and secure.
- **Intrusion Detection/Prevention Systems (IDS/IPS):** Verify that IDS/IPS are in place and actively monitoring for threats.

Access Controls:

- **User Access:** Ensure strict access controls and authentication mechanisms are in place (e.g., multi-factor authentication).
- **Segmentation:** Check for network segmentation to isolate critical systems and limit the spread of potential attacks.

4. Network Monitoring and Management

Monitoring Tools:

- **Real-Time Monitoring:** Use tools like PRTG Network Monitor, SolarWinds Network Performance Monitor, or Zabbix to continuously monitor network performance and health.
- **Alerts and Notifications:** Set up alerts for unusual activity, performance degradation, or security threats.

Traffic Analysis:

- **Traffic Patterns:** Analyze network traffic to identify bottlenecks, high-usage periods, and potential performance issues.
- **Quality of Service (QoS):** Implement QoS policies to prioritize critical traffic and ensure consistent performance.

5. Testing and Audits

Regular Audits:

- **Internal Audits:** Conduct regular internal audits to review network configurations, security measures, and performance metrics.
- **External Audits:** Engage third-party experts to perform comprehensive network audits and identify potential weaknesses.

Stress Testing:

- **Load Testing:** Use stress testing tools to simulate high traffic volumes and identify potential points of failure.
- **Failover Testing:** Regularly test failover mechanisms to ensure that redundancy systems work as expected.

6. Capacity Planning and Scalability

Current Capacity:

- **Utilization Metrics:** Monitor current network utilization to ensure it operates within safe limits without being overburdened.
- **Future Growth:** Plan for future capacity needs based on projected business growth and increased data demands.

Scalability:

- **Modular Design:** Ensure the network is designed in a modular fashion, allowing for easy expansion and upgrades.
- **Cloud Integration:** Consider hybrid solutions that leverage cloud infrastructure for scalability and flexibility.

7. Documentation and Training

Documentation:

- **Network Documentation:** Maintain detailed and up-to-date documentation of network configurations, policies, and procedures.
- **Change Management:** Implement a change management process to document and review changes to the network infrastructure.

Training:

- **Staff Training:** Ensure IT staff are well-trained in network management, security protocols, and the latest technologies.
- **User Education:** Educate end-users about best practices for network usage and security.

Tools for Assessing Network Infrastructure

- **Network Performance Monitoring:** SolarWinds NPM, PRTG Network Monitor, Nagios
- **Bandwidth and Speed Testing:** Speedtest by Ookla, iPerf, NetFlow Analyzer
- **Security Assessment:** Nessus, Wireshark, Snort
- **Documentation and Mapping:** Microsoft Visio, NetBrain, Lucidchart

By systematically evaluating these aspects, businesses can determine the robustness and reliability of their network infrastructure, identifying areas for improvement and ensuring that the network can support current and future business needs effectively.

How robust and reliable is your network infrastructure?

Are there any network limitations or challenges that could impact AI implementation?

Network security in a business is typically handled through a combination of policies, practices, and technologies designed to protect the business's digital assets from various threats. Here are the key components and best practices for managing network security in a business:

1. Firewalls

- **Hardware Firewalls:** Physical devices placed between the business's internal network and the external internet to filter and control incoming and outgoing traffic.
- **Software Firewalls:** Installed on individual computers or servers to provide an additional layer of security.

2. Antivirus and Anti-Malware Software

- **Comprehensive Protection:** Install and regularly update antivirus and anti-malware software on all devices to detect and remove malicious software.
- **Regular Scans:** Schedule regular system scans to identify and mitigate threats.

3. Secure Wi-Fi Networks

- **Strong Encryption:** Use WPA3 (or at least WPA2) encryption for wireless networks to prevent unauthorized access.
- **Hidden SSID:** Optionally, hide the SSID of the Wi-Fi network to make it less visible to outsiders.
- **Guest Networks:** Set up separate guest networks for visitors to limit access to the main business network.

4. Access Controls

- **User Authentication:** Implement strong password policies and multi-factor authentication (MFA) to verify user identities.
- **Role-Based Access Control (RBAC):** Restrict access to sensitive information based on the user's role within the organization.
- **Regular Reviews:** Periodically review user access rights and remove unnecessary privileges.

5. Data Encryption

- **Encryption at Rest:** Encrypt sensitive data stored on servers, databases, and storage devices.
- **Encryption in Transit:** Use SSL/TLS encryption for data transmitted over the internet to protect it from interception.

6. Network Monitoring

- **Intrusion Detection Systems (IDS):** Deploy IDS to monitor network traffic for suspicious activity and potential security breaches.

- **Logging and Alerts:** Maintain logs of network activity and set up alerts for unusual behavior to enable quick response to potential threats.

7. Regular Software Updates and Patch Management

- **Automatic Updates:** Enable automatic updates for operating systems and applications to ensure that security patches are applied promptly.
- **Patch Management:** Implement a patch management process to regularly update software and firmware on all devices.

8. Secure Remote Access

- **Virtual Private Network (VPN):** Use VPNs to secure remote access to the business network for employees working from home or on the go.
- **Remote Desktop Security:** Ensure remote desktop services are secure by using strong passwords, MFA, and limiting access to trusted IP addresses.

9. Employee Training and Awareness

- **Security Training:** Provide regular training sessions for employees on security best practices, phishing awareness, and safe internet usage.
- **Security Policies:** Develop and enforce security policies that outline acceptable use of company resources, data handling procedures, and incident reporting.

10. Backup and Recovery

- **Regular Backups:** Implement regular backups of critical data to protect against data loss due to cyber-attacks, hardware failures, or other disasters.
- **Off-Site Storage:** Store backups off-site or in the cloud to ensure data can be recovered even if the main business location is compromised.
- **Disaster Recovery Plan:** Develop and test a disaster recovery plan to ensure the business can quickly resume operations after a security incident.

11. Physical Security

- **Secure Access:** Restrict physical access to network infrastructure, servers, and critical devices to authorized personnel only.
- **Security Cameras:** Install security cameras and access controls to monitor and secure sensitive areas.

12. Use of Security Tools and Services

- **Security Suites:** Utilize comprehensive security suites that provide multiple layers of protection, such as firewall, antivirus, anti-malware, and intrusion prevention.
- **Managed Security Services:** Consider outsourcing to managed security service providers (MSSPs) for expert management and monitoring of network security.

Tools and Solutions for Network Security in Businesses

- **Firewall Solutions:** Cisco ASA, Fortinet FortiGate, SonicWall
- **Antivirus/Anti-Malware Software:** Norton Small Business, Bitdefender, Kaspersky Small Business
- **VPN Solutions:** NordVPN, ExpressVPN, Cisco AnyConnect
- **Network Monitoring Tools:** SolarWinds, PRTG Network Monitor, Nagios
- **Backup Solutions:** Backblaze, Carbonite, Acronis

By implementing these security measures and continuously monitoring and updating security practices, businesses can protect their networks and data from a wide range of cyber threats, ensuring the safety and integrity of their digital assets.

How is network security managed?

Technology Stack: Technology Integration

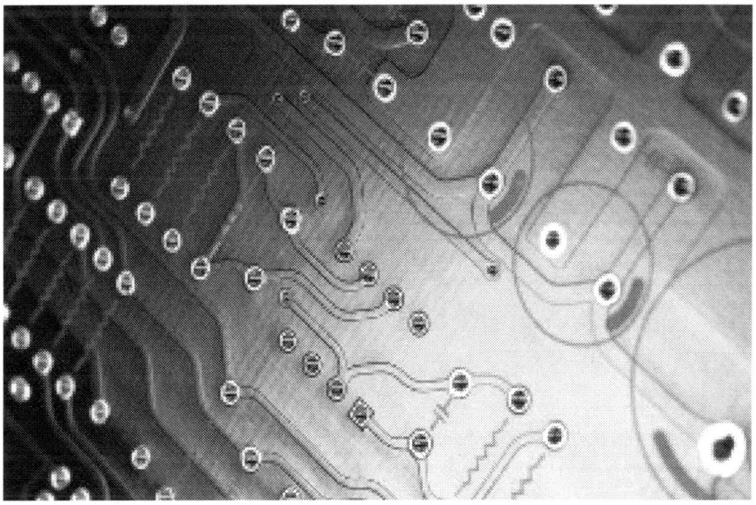

Determining how well your existing hardware and software systems integrate involves evaluating several key aspects of your IT infrastructure, including compatibility, performance, user experience, and maintenance. Here are steps to assess the integration of your systems:

1. Inventory and Documentation

Hardware and Software Inventory:

- **Catalog Systems:** Create a comprehensive inventory of all hardware and software systems in use, including versions, configurations, and interdependencies.
- **Documentation Review:** Ensure all system documentation is up to date, including network diagrams, system architecture, and data flow maps.

2. Compatibility Analysis

System Compatibility:

- **Vendor Specifications:** Check vendor documentation for compatibility information and recommended configurations.
- **Standards Compliance:** Verify that systems adhere to industry standards and protocols (e.g., TCP/IP, HTTP, SSL/TLS) to ensure interoperability.

Integration Points:

- **APIs and Interfaces:** Evaluate the availability and functionality of APIs, connectors, and interfaces that facilitate integration between systems.
- **Middleware:** Assess the role of middleware in your integration strategy and ensure it effectively connects disparate systems.

3. Performance and Reliability Testing

Performance Metrics:

- **Data Flow Efficiency:** Measure the speed and reliability of data transfer between systems. Tools like iPerf can help test network performance.
- **System Load Testing:** Use stress testing tools to evaluate how well integrated systems handle high loads and concurrent requests.

Reliability:

- **Uptime and Downtime:** Monitor system uptime and downtime to assess reliability. Tools like Nagios or SolarWinds can help track these metrics.
- **Error Rates:** Track the frequency and types of errors that occur during system interactions to identify potential integration issues.

4. User Experience and Feedback

User Surveys:

- **Feedback Collection:** Survey end-users to gather feedback on their experiences with system integration, focusing on ease of use, performance, and reliability.
- **Usability Issues:** Identify common usability issues and bottlenecks that users encounter when interacting with integrated systems.

Support Tickets:

- **Help Desk Analysis:** Review support tickets and incident reports related to system integration issues to identify recurring problems and areas for improvement.

5. Security and Compliance

Security Assessments:

- **Vulnerability Scans:** Conduct regular security assessments and vulnerability scans to ensure integrated systems are secure.
- **Access Controls:** Verify that access controls are properly configured to protect data and system integrity across integrated platforms.

Compliance Audits:

- **Regulatory Requirements:** Ensure that integrated systems comply with relevant regulatory requirements and industry standards (e.g., GDPR, HIPAA).

6. Data Consistency and Integrity

Data Quality Checks:

- **Data Validation:** Implement data validation rules and checks to ensure data consistency and integrity across integrated systems.
- **Data Reconciliation:** Perform regular data reconciliation between systems to identify and resolve discrepancies.

7. Integration Tools and Technologies

Integration Platforms:

- **iPaaS Solutions:** Evaluate the effectiveness of your Integration Platform as a Service (iPaaS) solutions, such as MuleSoft, Dell Boomi, or Talend.
- **ETL Tools:** Assess the performance of Extract, Transform, Load (ETL) tools like Informatica, Microsoft SSIS, or Apache Nifi in managing data integration tasks.

8. Scalability and Flexibility

Scalability Testing:

- **Capacity Planning:** Evaluate how well your integrated systems can scale to handle increased data volumes and user loads.
- **Future Proofing:** Ensure systems are flexible and can adapt to new technologies and business requirements.

9. Continuous Monitoring and Improvement

Monitoring Tools:

- **Real-Time Monitoring:** Implement real-time monitoring tools to continuously track the performance and health of integrated systems.
- **Automated Alerts:** Set up automated alerts to notify IT staff of integration issues or system failures.

Continuous Improvement:

- **Regular Reviews:** Conduct regular reviews and updates of integration processes and technologies to incorporate feedback and address emerging needs.
- **Training and Development:** Provide ongoing training for IT staff on integration tools and best practices.

Tools for Assessing Integration

- **Network Monitoring:** SolarWinds, PRTG Network Monitor, Nagios
- **Performance Testing:** iPerf, JMeter, LoadRunner
- **User Feedback:** SurveyMonkey, Google Forms
- **Data Quality:** Talend Data Quality, Informatica Data Quality
- **Security Assessment:** Nessus, Qualys, OpenVAS
- **Integration Platforms:** MuleSoft, Dell Boomi, Talend

By systematically evaluating these aspects, businesses can gain a comprehensive understanding of how well their hardware and software systems integrate and identify areas for improvement to enhance overall efficiency and performance.

How well do your existing hardware and software systems integrate?

Are there any interoperability issues?

What tools or strategies do you use to ensure seamless integration?

Skill Set: AI Knowledge and Experience

Do you have in-house experts with knowledge of AI Technologies?

What is the level of AI expertise among your staff (beginner, intermediate, advanced)?

Are there specific team members dedicated to AI projects?

Skill Set: Training and Development

What training programs are available for staff to learn about AI?

How often do you provide AI-related training and upskilling opportunities?

Are there partnerships with educational institutions or training providers for AI education?

Skill Set: Hiring and Recruitment

Hiring AI talent can be challenging due to the high demand and specialized skills required. Here are some effective strategies for attracting and retaining top AI talent:

1. Define Clear Roles and Requirements

- **Job Descriptions:** Write clear and detailed job descriptions that specify the required skills, experience, and qualifications. Differentiate between roles such as data scientist, machine learning engineer, AI researcher, and data engineer.
- **Specific Skills:** Highlight specific technical skills (e.g., Python, TensorFlow, PyTorch), domain expertise (e.g., natural language processing, computer vision), and soft skills (e.g., problem-solving, communication).

2. Leverage Multiple Recruitment Channels

- **Online Job Portals:** Use job boards specialized in tech and AI roles, such as LinkedIn, Glassdoor, Indeed, and AI-specific job boards like AI Jobs.
- **University Partnerships:** Establish partnerships with universities and participate in career fairs, internships, and cooperative education programs to attract fresh graduates.
- **Professional Networks:** Utilize professional networks and online communities such as GitHub, Kaggle, and AI-related forums to connect with AI professionals.

3. Offer Competitive Compensation and Benefits

- **Salary and Bonuses:** Offer competitive salaries, performance bonuses, and incentives to attract top talent. Use salary benchmarking tools to stay competitive.
- **Equity and Stock Options:** Provide equity or stock options to make the offer more attractive, especially for startup environments.
- **Comprehensive Benefits:** Include health insurance, retirement plans, flexible working hours, and remote work options.

4. Create a Strong Employer Brand

- **Company Culture:** Promote a culture of innovation, collaboration, and continuous learning. Highlight your company's commitment to AI and technology advancement.
- **Showcase Projects:** Publicize your AI projects, research, and achievements to demonstrate the exciting and impactful work potential hires can expect.
- **Employee Testimonials:** Share testimonials from current employees about their experiences and growth opportunities within the company.

5. Focus on Professional Development

- **Training Programs:** Offer ongoing training and development programs, including access to courses, certifications, and conferences.
- **Mentorship:** Establish mentorship programs where new hires can learn from experienced AI professionals within the organization.
- **Research Opportunities:** Encourage and support employees to engage in research, publish papers, and attend industry conferences.

6. Utilize AI Recruitment Tools

- **AI-Powered Recruitment:** Use AI-powered recruitment tools to streamline the hiring process, such as resume screening, candidate matching, and interview scheduling.
- **Assessment Platforms:** Implement technical assessment platforms like HackerRank, Codility, or LeetCode to evaluate candidates' coding and problem-solving skills.

7. Build a Diverse Team

- **Inclusive Hiring Practices:** Promote diversity and inclusion in your hiring process to attract a wide range of talent with different perspectives and ideas.
- **Global Recruitment:** Consider hiring remote AI talent from different geographical regions to tap into a broader talent pool.

8. Engage with the AI Community

- **Conferences and Meetups:** Participate in and sponsor AI conferences, workshops, hackathons, and meetups to network with AI professionals and showcase your company.
- **Open-Source Contributions:** Encourage and support employees to contribute to open-source AI projects, which can help attract talent who are active in the AI community.

9. Internal Talent Development

- **Upskilling Programs:** Identify potential talent within your organization and provide upskilling opportunities in AI and machine learning.
- **Career Pathways:** Create clear career pathways for employees to transition into AI roles, supported by training and development programs.

10. Recruitment Agencies and Headhunters

- **Specialized Agencies:** Partner with recruitment agencies and headhunters specialized in tech and AI talent to help identify and attract candidates.

Summary

To effectively hire AI talent, companies should focus on clear role definitions, competitive compensation, professional development opportunities, strong employer branding, diverse and inclusive hiring practices, engagement with the AI community, and leveraging modern recruitment tools and platforms. Combining these strategies will help attract and retain top AI professionals in a competitive job market.

Do you have a strategy for hiring AI talent?

When hiring for AI-related positions, it's important to look for a combination of technical qualifications, practical experience, and soft skills. Here are the key qualifications and skills to consider:

1. Educational Background

- **Degree in Relevant Field:** Look for candidates with degrees in computer science, data science, artificial intelligence, machine learning, statistics, mathematics, or related fields. Advanced degrees (Master's or PhD) are often preferred for research-intensive roles.

2. Technical Skills

Programming Languages:

- **Python:** The most commonly used language in AI and machine learning.
- **R:** Often used for statistical analysis and data visualization.
- **Other Languages:** Knowledge of Java, C++, Julia, or Scala can also be beneficial.

Machine Learning Frameworks and Libraries:

- **TensorFlow and Keras:** Widely used for deep learning projects.
- **PyTorch:** Popular for research and production in deep learning.
- **Scikit-Learn:** Essential for traditional machine learning tasks.
- **Other Tools:** Experience with libraries such as XGBoost, LightGBM, and OpenCV.

Data Handling and Analysis:

- **SQL:** For database querying and manipulation.
- **Pandas and NumPy:** For data manipulation and numerical computations.
- **Big Data Technologies:** Knowledge of Hadoop, Spark, or Apache Kafka for handling large datasets.

Statistical and Mathematical Skills:

- **Probability and Statistics:** Essential for understanding and applying machine learning algorithms.
- **Linear Algebra and Calculus:** Fundamental for developing and understanding AI models.

3. Domain Knowledge

- **Industry-Specific Knowledge:** Experience in the specific industry (e.g., healthcare, finance, retail) can be valuable for applying AI solutions effectively.
- **Understanding of AI Ethics:** Awareness of ethical considerations and biases in AI is increasingly important.

4. Experience and Projects

- **Practical Experience:** Hands-on experience with AI and machine learning projects, whether through previous job roles, internships, or personal projects.
- **Portfolio of Work:** A strong portfolio showcasing completed AI projects, contributions to open-source projects, or published research papers.
- **Competitions and Hackathons:** Participation in AI competitions (e.g., Kaggle) can demonstrate practical skills and problem-solving abilities.

5. Soft Skills

- **Problem-Solving:** Ability to approach complex problems methodically and develop innovative solutions.
- **Analytical Thinking:** Strong analytical skills to interpret data, identify patterns, and draw meaningful conclusions.
- **Communication:** Ability to explain complex technical concepts to non-technical stakeholders clearly and effectively.
- **Collaboration:** Experience working in cross-functional teams and collaborating with other departments.

6. Research and Development Skills

- **Algorithm Development:** Experience in developing and implementing new algorithms.
- **Research Experience:** For research-oriented roles, a track record of publishing papers in reputable journals or conferences.
- **Continuous Learning:** Demonstrated commitment to staying updated with the latest advancements in AI and machine learning.

7. Specializations (Depending on Role)

Data Scientist:

- **Data Cleaning and Preprocessing:** Skills in preparing and cleaning data for analysis.
- **Exploratory Data Analysis (EDA):** Proficiency in visualizing and exploring data to uncover insights.

Machine Learning Engineer:

- **Model Deployment:** Experience with deploying models to production environments.
- **Software Engineering Practices:** Knowledge of software development best practices, version control (e.g., Git), and DevOps tools.

AI Researcher:

- **Theoretical Knowledge:** Deep understanding of machine learning theories and concepts.
- **Experimental Skills:** Ability to design and conduct experiments to test new hypotheses.

Data Engineer:

- **ETL Processes:** Experience in designing and managing ETL (Extract, Transform, Load) pipelines.
- **Data Warehousing:** Knowledge of data warehousing solutions (e.g., Redshift, BigQuery).

Computer Vision Specialist:

- **Image Processing:** Skills in image processing and computer vision techniques.
- **Frameworks:** Experience with frameworks like OpenCV and tools like YOLO or Mask R-CNN.

Natural Language Processing (NLP) Expert:

- **Text Analysis:** Experience with NLP techniques and tools such as NLTK, spaCy, or Hugging Face.
- **Language Models:** Familiarity with transformer models (e.g., BERT, GPT).

Summary

When hiring for AI-related positions, focus on a mix of educational qualifications, technical skills, domain knowledge, practical experience, and soft skills. The exact combination will depend on the specific role and the needs of your organization. By targeting these key areas, you can find candidates who are well-equipped to contribute to your AI initiatives.

What qualifications and skills do you look for when hiring for AI-related positions?

Attracting and retaining AI talent presents several challenges due to the high demand and specialized nature of the field. Here are some key challenges and strategies to address them:

Challenges in Attracting AI Talent

1. **High Demand and Competition:**

   - **Challenge:** The demand for AI professionals far exceeds the supply, leading to intense competition among companies.
   - **Strategy:** Offer competitive salaries, benefits, and opportunities for growth to stand out in the market.
2. **Specialized Skill Set:**
   - **Challenge:** AI roles require a unique combination of technical skills, domain knowledge, and experience, making it difficult to find qualified candidates.
   - **Strategy:** Clearly define the skills and qualifications needed and consider training programs to develop these skills internally.
3. **Awareness and Branding:**
   - **Challenge:** Smaller companies or those not traditionally known for tech may struggle to attract AI talent.
   - **Strategy:** Build a strong employer brand by showcasing your AI projects, innovation, and company culture through various channels.
4. **Geographic Limitations:**
   - **Challenge:** AI talent may be concentrated in specific geographic regions, limiting the pool of candidates available locally.
   - **Strategy:** Offer remote work options and create a flexible work environment to attract talent from a broader geographic area.
5. **Lengthy Recruitment Process:**
   - **Challenge:** Lengthy and cumbersome recruitment processes can deter candidates, especially in a competitive market.
   - **Strategy:** Streamline the recruitment process to be efficient and responsive, while maintaining thoroughness in evaluating candidates.

Challenges in Retaining AI Talent

1. **High Expectations and Burnout:**
   - **Challenge:** AI professionals often face high expectations and heavy workloads, leading to burnout.
   - **Strategy:** Promote a healthy work-life balance, provide adequate resources, and ensure reasonable workloads to prevent burnout.
2. **Continuous Learning and Development:**
   - **Challenge:** AI is a rapidly evolving field, and professionals need continuous learning opportunities to stay current.
   - **Strategy:** Offer ongoing training, professional development, and opportunities to attend conferences and workshops.
3. **Career Advancement Opportunities:**
   - **Challenge:** AI talent may leave if they do not see clear career advancement opportunities within the company.
   - **Strategy:** Create clear career paths, offer promotions and new challenges, and provide leadership opportunities.
4. **Company Culture and Work Environment:**
   - **Challenge:** A poor company culture or unengaging work environment can lead to high turnover.

  - **Strategy:** Foster a positive, inclusive, and collaborative culture that values innovation and employee contributions.
5. **Competitive Offers from Other Companies:**
  - **Challenge:** AI talent frequently receives lucrative offers from other companies, leading to high turnover.
  - **Strategy:** Conduct regular market salary reviews and offer competitive compensation packages, including bonuses and stock options.
6. **Recognition and Reward:**
  - **Challenge:** Lack of recognition and reward for their contributions can demotivate AI professionals.
  - **Strategy:** Implement recognition programs to celebrate achievements and reward exceptional performance.

Strategies to Overcome Challenges

1. **Engage in University Partnerships:**
  - Collaborate with universities to create internship programs, sponsor AI research, and participate in career fairs to attract emerging talent.
2. **Build a Diverse Talent Pipeline:**
  - Focus on diversity and inclusion in your hiring practices to broaden your talent pool and create a more innovative team.
3. **Leverage Professional Networks and Communities:**
  - Participate in AI conferences, workshops, and online communities to connect with potential candidates and raise your company's profile.
4. **Offer Meaningful and Challenging Work:**
  - Ensure that AI professionals are engaged in meaningful projects that align with their interests and expertise, providing them with intellectual challenges and a sense of purpose.
5. **Create a Supportive Onboarding Process:**
  - Implement a comprehensive onboarding process that helps new hires integrate smoothly into the company and understand their roles and expectations.
6. **Provide Mentorship and Coaching:**
  - Establish mentorship programs where experienced professionals can guide and support newer employees, fostering a culture of continuous learning and growth.

By addressing these challenges and implementing effective strategies, businesses can attract and retain top AI talent, ensuring the success of their AI initiatives and maintaining a competitive edge in the market.

Are there challenges in attracting and retaining AI talent?

Skill Set: Team Collaboration

How well do different teams (IT, data science, business units, etc.) collaborate on AI projects?

Are there any communication barriers or silos that need to be addressed?

How are cross-functional AI initiatives managed?

Business Alignment: Strategic Objectives

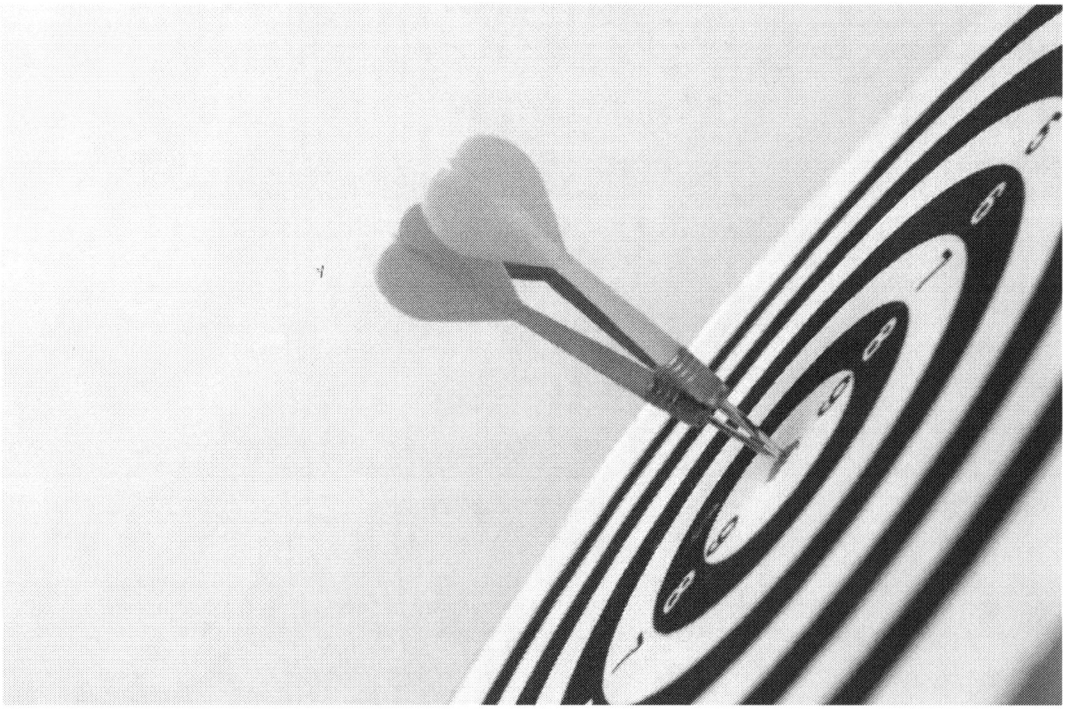

Aligning AI initiatives with the overall business strategy ensures that AI projects contribute to the company's goals and deliver tangible business value. Here are some key ways to achieve this alignment:

1. Define Clear Business Objectives

Identify Business Goals:

- **Strategic Priorities:** Understand the company's strategic priorities and long-term goals. Determine how AI can support these objectives, whether it's improving efficiency, enhancing customer experience, driving innovation, or creating new revenue streams.
- **Problem Areas:** Identify specific business problems or areas where AI can have the most significant impact. Focus on pain points that, when addressed, can deliver substantial benefits.

2. Develop a Comprehensive AI Strategy

Create an AI Roadmap:

- **Short-term and Long-term Goals:** Develop a clear roadmap that outlines short-term and long-term AI initiatives. Ensure these initiatives align with the company's overall business strategy.
- **Milestones and KPIs:** Establish milestones and key performance indicators (KPIs) to measure the success of AI projects and their alignment with business objectives.

3. Involve Key Stakeholders

Cross-Functional Collaboration:

- **Engage Leadership:** Involve senior leadership in AI strategy development to ensure alignment with business priorities and secure buy-in.
- **Interdepartmental Collaboration:** Foster collaboration between AI teams and other departments (e.g., marketing, sales, operations) to ensure AI projects address real business needs.

4. Focus on High-Impact Use Cases

Prioritize AI Projects:

- **Value and Feasibility:** Evaluate potential AI projects based on their expected business value and feasibility. Prioritize projects that offer the highest return on investment (ROI).
- **Pilot Programs:** Start with pilot programs to test AI solutions in a controlled environment before scaling them across the organization.

5. Invest in the Right Technology and Talent

AI Infrastructure:

- **Scalable Infrastructure:** Invest in scalable and flexible AI infrastructure that can grow with the company's needs. This includes data storage, computing power, and AI development platforms.
- **Integration Capabilities:** Ensure AI solutions can be integrated seamlessly with existing systems and workflows.

Talent Acquisition:

- **Skilled Workforce:** Hire or train employees with the necessary AI skills and expertise. Encourage continuous learning and development to keep up with advancements in AI technology.
- **Interdisciplinary Teams:** Build interdisciplinary teams that combine technical AI skills with business acumen to bridge the gap between AI development and business strategy.

6. Foster a Data-Driven Culture

Data Governance:

- **Data Quality:** Implement strong data governance practices to ensure the quality, accuracy, and reliability of data used in AI initiatives.
- **Data Accessibility:** Ensure relevant data is accessible to AI teams while maintaining data security and privacy.

Encourage Data-Driven Decision-Making:

- **Analytical Mindset:** Promote an analytical mindset across the organization, encouraging employees to leverage data and AI insights in their decision-making processes.
- **Transparency and Communication:** Communicate the value and impact of AI projects to all levels of the organization to build trust and support for AI initiatives.

7. Monitor and Measure Outcomes

Performance Tracking:

- **Regular Reviews:** Conduct regular reviews of AI projects to assess their progress and impact on business objectives.
- **KPIs and Metrics:** Use predefined KPIs and metrics to measure the success of AI initiatives and their contribution to the overall business strategy.

Continuous Improvement:

- **Feedback Loops:** Establish feedback loops to gather insights from stakeholders and end-users. Use this feedback to refine and improve AI solutions.

- **Agile Approach:** Adopt an agile approach to AI development, allowing for iterative improvements and adjustments based on real-world performance and changing business needs.

8. Ensure Ethical and Responsible AI Use

Ethical Guidelines:

- **AI Ethics:** Develop and implement ethical guidelines for AI use to ensure that AI initiatives align with the company's values and societal expectations.
- **Bias and Fairness:** Address issues of bias and fairness in AI models to build trust and ensure ethical use of AI technology.

Compliance and Regulation:

- **Regulatory Compliance:** Ensure all AI initiatives comply with relevant regulations and industry standards to avoid legal and reputational risks.

Summary

Aligning AI initiatives with overall business strategy involves defining clear business objectives, developing a comprehensive AI strategy, involving key stakeholders, focusing on high-impact use cases, investing in the right technology and talent, fostering a data-driven culture, monitoring outcomes, and ensuring ethical and responsible AI use. By following these steps, businesses can maximize the value of their AI investments and ensure that AI initiatives drive meaningful business outcomes

How are AI initiatives aligned with overall business strategy?

What specific business goals are you aiming to achieve with AI implementation?

Measuring the success of AI initiatives involves tracking key performance indicators (KPIs) that align with both the specific goals of the AI project and the overall business objectives. Here are some important KPIs to consider:

Technical Performance KPIs

1. **Model Accuracy:**
    - **Definition:** Measures how often the AI model makes correct predictions.
    - **Example Metrics:** Accuracy, precision, recall, F1 score, AUC-ROC.
2. **Model Training Time:**
    - **Definition:** The time it takes to train the AI model from start to finish.
    - **Example Metrics:** Hours or days to train a model.
3. **Inference Time:**
    - **Definition:** The time it takes for the model to make a prediction once it receives new data.
    - **Example Metrics:** Milliseconds or seconds per inference.
4. **Data Quality:**
    - **Definition:** Measures the quality and integrity of the data used for training and testing the model.
    - **Example Metrics:** Percentage of missing values, data accuracy, data consistency.
5. **Model Scalability:**
    - **Definition:** The ability of the model to handle increasing amounts of data or users without performance degradation.
    - **Example Metrics:** Throughput, latency under load.

Business Impact KPIs

6. **Return on Investment (ROI):**
    - **Definition:** Measures the financial return generated by the AI initiative compared to its cost.
    - **Example Metrics:** ROI percentage, net profit from AI implementation.
7. **Cost Savings:**
    - **Definition:** The reduction in costs as a result of implementing AI solutions.
    - **Example Metrics:** Cost reduction in operations, labor cost savings.
8. **Revenue Growth:**
    - **Definition:** Increase in revenue attributed to AI-driven improvements.
    - **Example Metrics:** Additional revenue generated, percentage increase in sales.
9. **Customer Satisfaction:**
    - **Definition:** Measures how AI impacts customer experience and satisfaction.
    - **Example Metrics:** Net Promoter Score (NPS), Customer Satisfaction Score (CSAT), reduction in customer complaints.
10. **Process Efficiency:**
    - **Definition:** Improvement in the efficiency of business processes due to AI automation and optimization.
    - **Example Metrics:** Reduction in process cycle time, increase in throughput, reduction in error rates.

Operational KPIs

11. **Adoption Rate:**
    - **Definition:** The rate at which the AI solution is adopted by users within the organization.
    - **Example Metrics:** Number of active users, percentage of processes using AI.
12. **User Engagement:**
    - **Definition:** Measures how engaged users are with the AI system.
    - **Example Metrics:** Frequency of use, time spent using the AI system, user feedback ratings.
13. **Compliance and Risk Reduction:**
    - **Definition:** The extent to which AI helps in reducing compliance risks and adhering to regulations.
    - **Example Metrics:** Number of compliance issues detected and resolved, reduction in regulatory fines.

AI Model Maintenance KPIs

14. **Model Drift:**
    - **Definition:** Measures how much the model's performance degrades over time as new data is introduced.
    - **Example Metrics:** Decrease in accuracy or other performance metrics over time.
15. **Retraining Frequency:**
    - **Definition:** The frequency at which the AI model needs to be retrained to maintain performance.
    - **Example Metrics:** Number of retraining cycles per year.
16. **Explainability and Transparency:**
    - **Definition:** Measures the ability to understand and interpret the AI model's decisions.
    - **Example Metrics:** Percentage of decisions that can be explained, user trust in AI decisions.

Innovation and Research KPIs

17. **Research and Development Output:**
    - **Definition:** The contributions of AI initiatives to research and development.
    - **Example Metrics:** Number of patents filed, research papers published, innovative solutions developed.
18. **Time to Market:**
    - **Definition:** The time it takes to develop and deploy AI solutions from inception to production.
    - **Example Metrics:** Weeks or months from project start to deployment.

Summary

By tracking these KPIs, businesses can effectively measure the success of their AI initiatives and ensure they are aligned with strategic objectives, delivering tangible benefits, and continuously improving over time.

Are there any key performance indicators (KPIs) to measure AI success?

Business Alignment: Leadership and Support

How supportive is the leadership team of AI initiatives?

Is there a designated AI champion or sponsor within the organization?

How are AI projects prioritized and funded?

Business Alignment: Use Cases and Applications

What are the primary use cases for AI in your business?

How do these use cases address current business challenges or opportunities?

Are there any successful AI projects or pilots that can be referenced?

Business Alignment: Risk Management

Implementing AI can bring substantial benefits, but it also comes with several potential risks that businesses need to be aware of and manage carefully. Here are some of the key risks associated with AI implementation:

1. Ethical and Bias Concerns

Bias in AI Models:

- **Risk:** AI models can inherit biases present in the training data, leading to unfair or discriminatory outcomes.
- **Mitigation:** Ensure diverse and representative training data, implement bias detection and mitigation techniques, and regularly audit AI systems for fairness.

Ethical Considerations:

- **Risk:** AI applications may raise ethical concerns, such as invasion of privacy, manipulation, or unintended harm.

- **Mitigation:** Develop and adhere to ethical guidelines for AI use, ensure transparency in AI decision-making, and engage in ethical reviews.

2. Data Privacy and Security

Data Breaches:

- **Risk:** AI systems often rely on large amounts of data, increasing the risk of data breaches and unauthorized access.
- **Mitigation:** Implement robust data security measures, including encryption, access controls, and regular security audits.

Privacy Violations:

- **Risk:** AI applications may inadvertently violate user privacy, especially in areas like surveillance and personal data analysis.
- **Mitigation:** Ensure compliance with data privacy regulations (e.g., GDPR, CCPA), and implement privacy-by-design principles.

3. Technical Challenges

Model Performance and Reliability:

- **Risk:** AI models may not perform as expected, leading to incorrect predictions or unreliable results.
- **Mitigation:** Rigorous testing and validation, continuous monitoring, and implementing fallback mechanisms.

Scalability Issues:

- **Risk:** AI systems may struggle to scale with increasing data volumes or user demands.
- **Mitigation:** Design scalable architectures, leverage cloud infrastructure, and optimize resource allocation.

4. Operational Risks

Integration Challenges:

- **Risk:** Difficulty integrating AI systems with existing IT infrastructure and workflows.
- **Mitigation:** Conduct thorough integration planning, use compatible technologies, and engage experienced integrators.

Maintenance and Updating:

- **Risk:** AI models require regular maintenance and updating to remain effective and accurate.

- **Mitigation:** Establish a robust model management framework, automate retraining processes, and monitor model performance continuously.

5. Workforce Impact

Job Displacement:

- **Risk:** AI implementation may lead to job displacement or changes in job roles, causing workforce disruption.
- **Mitigation:** Invest in employee reskilling and upskilling, communicate transparently with staff, and explore redeployment opportunities.

Resistance to Change:

- **Risk:** Employees may resist adopting AI technologies due to fear of job loss or unfamiliarity with new tools.
- **Mitigation:** Foster a culture of innovation, provide training and support, and involve employees in the AI adoption process.

6. Legal and Regulatory Risks

Compliance Issues:

- **Risk:** Failure to comply with relevant laws and regulations can result in legal penalties and reputational damage.
- **Mitigation:** Stay informed about regulatory requirements, conduct regular compliance audits, and engage legal experts.

Liability Concerns:

- **Risk:** Determining liability for AI decisions can be complex, especially in cases of harm or error.
- **Mitigation:** Clearly define accountability, establish robust oversight mechanisms, and ensure transparency in AI decision-making.

7. Strategic and Competitive Risks

Misalignment with Business Goals:

- **Risk:** AI projects that are not aligned with business objectives may fail to deliver expected benefits.
- **Mitigation:** Align AI initiatives with strategic goals, involve key stakeholders, and regularly review project progress.

Over-Reliance on AI:

- **Risk:** Over-reliance on AI can lead to complacency and reduced human oversight, increasing the risk of errors.
- **Mitigation:** Maintain a balanced approach, combine AI with human judgment, and ensure regular reviews.

8. Financial Risks

High Costs:

- **Risk:** AI projects can be costly, with expenses related to technology, talent, and ongoing maintenance.
- **Mitigation:** Conduct cost-benefit analyses, prioritize high-impact projects, and explore cost-effective solutions such as cloud services.

Return on Investment (ROI) Uncertainty:

- **Risk:** Uncertainty about the financial return from AI investments.
- **Mitigation:** Set clear ROI metrics, pilot projects to assess viability, and continuously measure and optimize performance.

Summary

By recognizing and addressing these risks through careful planning, robust governance, and continuous monitoring, businesses can mitigate potential issues and successfully leverage AI to achieve their strategic objectives.

What potential risks are associated with Ai implementation?

Identifying, assessing, and mitigating risks associated with AI implementation involves a structured approach that includes risk management frameworks, continuous monitoring, and proactive measures. Here are the steps to effectively manage AI-related risks:

Identifying Risks

1. **Stakeholder Engagement:**

- **Involvement:** Engage key stakeholders, including AI developers, business leaders, legal advisors, and end-users, to identify potential risks from various perspectives.
- **Workshops:** Conduct risk identification workshops to gather insights and concerns from different departments.

2. **Risk Identification Frameworks:**
   - **SWOT Analysis:** Use SWOT (Strengths, Weaknesses, Opportunities, Threats) analysis to identify internal and external risks.
   - **Risk Registers:** Maintain a risk register to document identified risks, their sources, and potential impacts.
3. **Data and Process Review:**
   - **Data Audit:** Perform a comprehensive audit of the data used for AI training and deployment to identify biases, privacy issues, and security vulnerabilities.
   - **Process Mapping:** Map out AI workflows and processes to identify points where risks may arise, such as data handling, model training, and decision-making.
4. **Scenario Analysis:**
   - **Hypothetical Scenarios:** Develop hypothetical scenarios to explore potential risks and their impacts. This includes worst-case scenarios to prepare for extreme situations.

Assessing Risks

1. **Risk Impact and Likelihood:**
   - **Impact Analysis:** Evaluate the potential impact of each risk on the business, considering factors such as financial loss, reputational damage, and operational disruption.
   - **Likelihood Assessment:** Assess the probability of each risk occurring based on historical data, expert judgment, and predictive modeling.
2. **Risk Prioritization:**
   - **Risk Matrix:** Use a risk matrix to prioritize risks based on their impact and likelihood. High-impact, high-likelihood risks should be addressed first.
   - **Risk Scoring:** Assign scores to risks to quantify their severity and facilitate comparison.
3. **Regulatory and Compliance Review:**
   - **Legal Assessment:** Review relevant regulations and compliance requirements to assess legal and regulatory risks associated with AI implementation.

Mitigating Risks

1. **Technical Measures:**
   - **Bias Mitigation:** Implement techniques to detect and mitigate bias in AI models, such as re-sampling data, using fairness-aware algorithms, and regular bias audits.
   - **Security Measures:** Strengthen data security through encryption, access controls, and regular security testing.
   - **Model Monitoring:** Continuously monitor AI model performance to detect and address model drift, inaccuracies, and anomalies.

2. **Policy and Governance:**
   - **Ethical Guidelines:** Develop and enforce ethical guidelines for AI use to ensure responsible and fair AI practices.
   - **Data Governance:** Implement robust data governance frameworks to ensure data quality, privacy, and compliance.
3. **Process Improvements:**
   - **Risk Management Plan:** Develop a comprehensive risk management plan that includes risk mitigation strategies, response plans, and contingency measures.
   - **Incident Response:** Establish incident response protocols to quickly address and mitigate the impact of AI-related issues.
4. **Training and Awareness:**
   - **Employee Training:** Provide training programs to enhance employees' understanding of AI risks, ethical considerations, and best practices.
   - **Stakeholder Communication:** Regularly communicate with stakeholders about risk management efforts, findings, and improvements.
5. **Continuous Improvement:**
   - **Feedback Loops:** Establish feedback loops to gather insights from AI users and stakeholders, using this feedback to improve AI systems and risk management practices.
   - **Regular Audits:** Conduct regular audits and reviews of AI systems to ensure ongoing compliance, performance, and risk mitigation.

Tools and Techniques for Risk Management

1. **Risk Management Software:** Use software tools like RiskWatch, LogicManager, or RSA Archer to manage risk identification, assessment, and mitigation processes.
2. **AI Monitoring Tools:** Implement AI-specific monitoring tools like IBM Watson OpenScale, Google Cloud AI Platform, or Azure Machine Learning to track model performance and detect issues.
3. **Data Quality Tools:** Utilize data quality and governance tools like Talend Data Quality, Informatica, or Collibra to ensure the integrity and reliability of data used in AI projects.

Summary

By systematically identifying, assessing, and mitigating risks associated with AI implementation, businesses can minimize potential negative impacts and ensure that AI initiatives contribute positively to their strategic objectives. This structured approach involves stakeholder engagement, robust risk management frameworks, continuous monitoring, and proactive measures to address risks effectively.

How are these risks identified, assessed and mitigated?

Contingency plans for AI project failures are essential to ensure that businesses can quickly respond to issues and mitigate their impact. Here are various types of contingency plans that can be put in place:

1. Technical Contingency Plans

Rollback Procedures:

- **Definition:** Processes for reverting to a previous stable state in case of AI system failure.
- **Implementation:** Regularly create and test backups of models, data, and system configurations. Have scripts or automated processes in place to revert to these backups quickly.

Redundancy and Failover Systems:

- **Definition:** Secondary systems that can take over in case the primary AI system fails.
- **Implementation:** Use redundant servers and storage and ensure failover mechanisms are in place to switch to backup systems seamlessly.

Manual Override Mechanisms:

- **Definition:** Allow human intervention to take control when AI systems fail or produce incorrect results.
- **Implementation:** Design systems with manual override capabilities and train staff on when and how to use these controls.

2. Operational Contingency Plans

Incident Response Teams:

- **Definition:** Designated teams responsible for managing AI system failures.
- **Implementation:** Form incident response teams with clear roles and responsibilities, and ensure they are trained in handling AI-related incidents.

Communication Plans:

- **Definition:** Protocols for communicating with stakeholders during an AI project failure.

- **Implementation:** Develop communication templates and channels to inform stakeholders, including employees, customers, and partners, about the issue and the steps being taken to resolve it.

Business Continuity Plans:

- **Definition:** Strategies to ensure that critical business operations continue despite AI system failures.
- **Implementation:** Identify critical business functions and develop plans to maintain them using alternative methods or systems.

3. Data and Model Contingency Plans

Data Integrity Checks:

- **Definition:** Regular validation of data quality and integrity to prevent AI failures caused by data issues.
- **Implementation:** Implement automated data validation processes and set up alerts for data anomalies.

Model Validation and Testing:

- **Definition:** Rigorous testing and validation of AI models before deployment to ensure they perform as expected.
- **Implementation:** Use cross-validation, stress testing, and scenario analysis to validate models. Regularly update and retrain models to maintain their accuracy.

Alternative Models:

- **Definition:** Backup models that can be deployed if the primary model fails.
- **Implementation:** Develop and maintain secondary models that can be quickly deployed in case of issues with the primary model.

4. Compliance and Legal Contingency Plans

Regulatory Compliance Checks:

- **Definition:** Ensure that AI systems comply with relevant regulations and standards to avoid legal issues.
- **Implementation:** Regularly review and update compliance measures and conduct audits to ensure adherence to regulations.

Legal Response Plans:

- **Definition:** Plans to address legal challenges arising from AI failures.

- **Implementation:** Have legal teams prepared to handle potential lawsuits or regulatory actions. Maintain documentation of AI development and deployment processes to support legal defense.

5. Financial Contingency Plans

Budget Allocations:

- **Definition:** Set aside funds specifically for addressing AI system failures and related contingencies.
- **Implementation:** Include contingency budgets in AI project planning to cover costs of recovery, legal fees, and additional resources.

Insurance Policies:

- **Definition:** Insurance coverage for risks associated with AI systems.
- **Implementation:** Obtain insurance policies that cover technology errors, data breaches, and business interruptions related to AI failures.

6. Human Resources Contingency Plans

Training and Development:

- **Definition:** Regular training for employees on handling AI system failures and manual processes.
- **Implementation:** Conduct regular training sessions and simulations to ensure employees are prepared to respond to AI-related issues.

Staff Augmentation:

- **Definition:** Plans to quickly bring in additional personnel or expertise in case of major AI system failures.
- **Implementation:** Maintain relationships with consulting firms and staffing agencies that can provide temporary or specialized support when needed.

7. Continuous Improvement and Monitoring

Monitoring and Alerts:

- **Definition:** Continuous monitoring of AI systems to detect and address issues proactively.
- **Implementation:** Use monitoring tools to track system performance, set up alerts for anomalies, and implement automated corrective actions.

Post-Mortem Analysis:

- **Definition:** Review and analyze AI failures to identify root causes and prevent recurrence.
- **Implementation:** Conduct thorough post-mortem analyses after each incident, document findings, and update processes and systems based on lessons learned.

Summary

By implementing these contingency plans, businesses can better prepare for and respond to AI project failures, minimizing their impact and ensuring continuity of operations. These plans should be regularly reviewed and updated to adapt to new challenges and technological advancements.

Are there contingency plans in place for AI project failures?

How is change in management handled in your organization?

Ensuring a smooth adoption of AI technologies during a change in management requires strategic planning, effective communication, and strong leadership. Here are some strategies to facilitate this transition:

1. Develop a Clear AI Vision and Strategy

Align AI with Business Goals:

- **Strategy:** Ensure that the AI vision aligns with the overall business goals and objectives. Communicate how AI will drive business value and improve operations.
- **Implementation:** Create a strategic roadmap for AI adoption that outlines short-term and long-term goals, key milestones, and success metrics.

2. Engage and Educate New Leadership

Comprehensive Onboarding:

- **Strategy:** Provide new management with a thorough understanding of the current AI initiatives, their objectives, progress, and expected outcomes.
- **Implementation:** Conduct detailed onboarding sessions, share documentation, and arrange meetings with key AI project stakeholders.

AI Training and Awareness:

- **Strategy:** Enhance new leaders' understanding of AI technologies, their potential, and best practices.
- **Implementation:** Offer training programs, workshops, and resources on AI fundamentals, use cases, and strategic benefits.

3. Foster a Culture of Collaboration and Innovation

Cross-Functional Teams:

- **Strategy:** Encourage collaboration between AI teams and other departments to ensure AI initiatives are integrated and supported across the organization.
- **Implementation:** Establish cross-functional teams that include members from IT, operations, marketing, finance, and other relevant areas.

Open Communication Channels:

- **Strategy:** Maintain open and transparent communication channels to keep everyone informed and engaged with AI initiatives.
- **Implementation:** Use regular updates, meetings, and internal newsletters to share progress, challenges, and success stories.

4. Establish Strong Governance and Oversight

AI Governance Framework:

- **Strategy:** Implement a robust governance framework to ensure AI projects are aligned with business objectives and ethical standards.
- **Implementation:** Define roles, responsibilities, and processes for overseeing AI initiatives, including data management, model validation, and compliance.

Regular Reviews and Audits:

- **Strategy:** Conduct regular reviews and audits to assess the performance, risks, and alignment of AI projects.
- **Implementation:** Schedule periodic reviews with new management to discuss AI project status, challenges, and opportunities for improvement.

5. Focus on Change Management

Change Management Plan:

- **Strategy:** Develop a comprehensive change management plan to guide the organization through the transition.
- **Implementation:** Include clear communication strategies, training programs, and support mechanisms to help employees adapt to AI technologies.

Change Champions:

- **Strategy:** Identify and empower change champions within the organization to advocate for AI adoption and support their peers.
- **Implementation:** Select individuals who are enthusiastic about AI and provide them with the resources and authority to drive change.

6. Demonstrate Quick Wins and Value

Pilot Projects:

- **Strategy:** Implement pilot AI projects that can demonstrate quick wins and tangible benefits to the organization.
- **Implementation:** Select pilot projects that address high-impact areas and have a clear path to success, then showcase the results to build momentum and support.

Success Metrics:

- **Strategy:** Define and track key performance indicators (KPIs) to measure the success and impact of AI initiatives.
- **Implementation:** Use KPIs such as cost savings, efficiency improvements, revenue growth, and customer satisfaction to demonstrate the value of AI projects.

7. Ensure Continuity and Knowledge Transfer

Documentation and Knowledge Sharing:

- **Strategy:** Maintain comprehensive documentation of AI projects, including objectives, methodologies, data sources, and results.
- **Implementation:** Create a centralized repository for documentation and encourage knowledge sharing through collaborative platforms and regular knowledge-sharing sessions.

Transition Plans:

- **Strategy:** Develop transition plans to ensure continuity of AI initiatives during management changes.

- **Implementation:** Identify key personnel responsible for AI projects and establish backup roles to ensure seamless transitions.

8. Secure Executive Sponsorship

Executive Support:

- **Strategy:** Ensure continuous support from senior executives to champion AI initiatives and allocate necessary resources.
- **Implementation:** Regularly update executives on AI progress and involve them in key decision-making processes.

Stakeholder Buy-In:

- **Strategy:** Engage key stakeholders across the organization to build consensus and support for AI adoption.
- **Implementation:** Conduct stakeholder meetings, workshops, and presentations to address concerns and highlight the strategic benefits of AI.

Summary

By implementing these strategies, organizations can ensure a smooth adoption of AI technologies during a change in management, fostering a culture of innovation, collaboration, and continuous improvement. These efforts will help align AI initiatives with business goals and demonstrate their value, securing long-term success and support.

What strategies are in place to ensure smooth adoption of AI technologies?

Handling resistance to change among employees is critical for the successful adoption of new technologies, including AI. Here are several strategies to effectively manage and mitigate resistance:

1. Communicate Clearly and Transparently

Open Communication Channels:

- **Strategy:** Establish open and transparent communication channels to keep employees informed about the changes and the reasons behind them.
- **Implementation:** Hold regular town hall meetings, send out newsletters, and use internal communication platforms to share updates and address concerns.

Articulate the Vision and Benefits:

- **Strategy:** Clearly articulate the vision, benefits, and positive impact of the changes on the organization and individual employees.
- **Implementation:** Create and share a compelling narrative that highlights how the change aligns with the company's goals and values.

2. Involve Employees in the Change Process

Engage Employees Early:

- **Strategy:** Involve employees early in the planning and implementation process to give them a sense of ownership and control.
- **Implementation:** Form cross-functional teams, seek input and feedback, and incorporate their suggestions into the change plan.

Empower Change Champions:

- **Strategy:** Identify and empower change champions within the organization to advocate for the change and support their peers.
- **Implementation:** Select respected and influential employees who are enthusiastic about the change and provide them with the resources and authority to drive it.

3. Provide Training and Support

Comprehensive Training Programs:

- **Strategy:** Offer comprehensive training programs to equip employees with the necessary skills and knowledge to adapt to the change.
- **Implementation:** Develop and deliver training sessions, workshops, and online courses tailored to different roles and levels of expertise.

Continuous Support:

- **Strategy:** Provide ongoing support to help employees navigate the change and address any challenges they may encounter.
- **Implementation:** Set up help desks, create FAQ resources, and offer one-on-one coaching or mentoring.

4. Address Concerns and Fears

Listen and Acknowledge:

- **Strategy:** Actively listen to employees' concerns and fears about the change and acknowledge their feelings.
- **Implementation:** Conduct surveys, focus groups, and individual meetings to gather feedback and demonstrate empathy and understanding.

Provide Reassurance:

- **Strategy:** Reassure employees about job security and the organization's commitment to their well-being.
- **Implementation:** Communicate any plans for reskilling, upskilling, and potential new opportunities that the change may bring.

5. Demonstrate Commitment and Leadership

Lead by Example:

- **Strategy:** Ensure that leaders and managers visibly support and participate in the change process.
- **Implementation:** Have leaders actively engage in training, communicate regularly about the change, and demonstrate their commitment through their actions.

Consistent Messaging:

- **Strategy:** Ensure that the messaging about the change is consistent across all levels of the organization.
- **Implementation:** Align communication from senior leaders, managers, and change champions to reinforce the key messages and vision.

6. Implement Change Gradually

Phased Approach:

- **Strategy:** Implement the change in phases to allow employees to adjust gradually and provide feedback along the way.
- **Implementation:** Start with pilot projects or small-scale implementations before rolling out the change organization-wide.

Monitor and Adjust:

- **Strategy:** Continuously monitor the progress of the change and be flexible in adjusting the approach as needed.
- **Implementation:** Use feedback, performance metrics, and regular reviews to identify areas for improvement and make necessary adjustments.

7. Celebrate Successes and Milestones

Recognize Achievements:

- **Strategy:** Celebrate successes and milestones to build momentum and demonstrate the positive impact of the change.
- **Implementation:** Acknowledge individual and team contributions through awards, recognition programs, and public commendations.

Share Success Stories:

- **Strategy:** Share success stories and case studies that highlight the benefits and positive outcomes of the change.
- **Implementation:** Use internal newsletters, meetings, and social media to showcase success stories and inspire others.

8. Create a Supportive Environment

Foster a Culture of Innovation:

- **Strategy:** Cultivate a culture that encourages innovation, experimentation, and continuous improvement.
- **Implementation:** Provide opportunities for employees to explore new ideas, take risks, and learn from failures.

Build Trust:

- **Strategy:** Build trust between management and employees by being transparent, fair, and responsive to their needs.
- **Implementation:** Establish regular feedback loops, act on employee input, and maintain open lines of communication.

Summary

By employing these strategies, organizations can effectively manage resistance to change and foster a positive, supportive environment that encourages employee engagement and buy-in. This will ultimately lead to a smoother transition and successful adoption of new technologies and processes.

How do you handle resistance to change among employees?

Conclusion

Based on your responses, what are the top three (3) areas that need improvement to enhance your AI readiness?

What immediate actions can be taken to address these areas?

Roadmaps & Timelines

6-Month Roadmap & Timeline

Name of AI Project: ______________________________

Month 1: Planning Phase

- ☐ **Define Objectives:** Clearly outline the goals of your AI Project
- ☐ **Set Budget:** Establish a budget for the project
- ☐ **Assemble Team:** Identify key team members and define their roles
- ☐ **Feasibility Study:** Conduct a study to evaluate the feasibility of the AI project
- ☐ **Select AI Use Cases:** Choose specific use cases that align with business objectives

Month 2: Development Phase – Data Preparation

- ☐ **Data Collection:** Gather relevant data from various sources
- ☐ **Data Cleaning:** Ensure data quality through cleaning and preprocessing
- ☐ **Select Tools and Platforms:** Choose the AI tools and platforms to be used

Month 3-4: Development Phase – Model Development

- ☐ **Model Development:** Develop and train AI models
- ☐ **Iterative Improvement:** Iterate on model development to improve accuracy and performance

Month 5: Development Phase – Pilot Testing

- ☐ **Pilot Testing:** Conduct initial tests of the AI Models
- ☐ **Feedback Collection:** Validate models with real-world data and gather feedback

Month 6: Deployment Phase – Initial Deployment

- ☐ **Integration:** Integrate AI models into existing systems
- ☐ **Scalability Planning:** Ensure systems can scale to meet increased demand
- ☐ **User Training:** Train end-users on how to interact with and utilize the AI System

Name of AI Project: ______________________________

Month	Activities	Start Date	Target Date	Completion Date
1	Planning Phase			
2	Data Preparation			
3-4	Model Development			
5	Pilot Testing			
6	Initial Deployment			

12-Month Roadmap & Timeline

Name of AI Project: ______________________________

Month 1-3: Planning Phase

- ☐ **Define Objectives:** Clearly outline the goals of your AI project
- ☐ **Set Budget:** Establish a budget for the project
- ☐ **Assemble Team:** Identify key team members and define their roles
- ☐ **Feasibility Study:** Conduct a study to evaluate the feasibility of the AI project
- ☐ **Select AI Use Cases:** Choose specific use cases that align with business objectives

Month 4-6: Development Phase – Data Preparation

- ☐ **Data Collection:** Gather relevant data from various sources
- ☐ **Data Cleaning:** Ensure data quality through cleaning and preprocessing
- ☐ **Select Tools and Platforms:** Choose the AI tools and platforms to be used

Month 7-8: Development Phase – Model Development

- ☐ **Model Development:** Develop and train AI models
- ☐ **Iterative Improvement:** Iterate on model development to improve accuracy and performance

Month 9: Development Phase – Pilot Testing

- ☐ **Pilot Testing:** Conduct initial tests of the AI models
- ☐ **Feedback Collection:** Validate models with real-world data and gather feedback

Month 10: Deployment Phase – Initial Deployment

- ☐ **Integration:** Integrate AI models into existing systems
- ☐ **Scalability Planning:** Ensure systems can scale to meet increased demand
- ☐ **User Training:** Train end-users on how to interact with and utilize the AI system

Month 11-12: Post-Deployment Phase

- ☐ **Feedback Loop:** Establish a feedback loop for continuous improvement
- ☐ **Performance Metrics:** Regularly review performance metrics
- ☐ **Periodic Audits:** Conduct periodic audits to ensure compliance and performance

Name of AI Project: ______________________________

Month	Activities	Start Date	Target Date	Completion Date
1-3	Planning Phase			
4-6	Data Preparation			
7-8	Model Development			
9	Pilot Testing			
10	Initial Deployment			
11-12	Post-Deployment			

AI Project Worksheets and Forms

Objectives and Goals Worksheet

Project Overview

- **Project Name:** ______________________________
- **Project Description:** ______________________________
- **Project Manager:** ______________________________
- **Team Members:** ______________________________
- **Start Date:** ______________________________
- **End Date:** ______________________________

Objectives

1. **Primary Objective:**
    - **Description:** ______________________________
    - **Success Metrics:** ______________________________
    - **Deadline:** ______________________________
2. **Secondary Objective:**
    - **Description:** ______________________________
    - **Success Metrics:** ______________________________
    - **Deadline:** ______________________________
3. **Additional Objectives:**
    - **Description:** ______________________________
    - **Success Metrics:** ______________________________
    - **Deadline:** ______________________________

Goals

Goal 1:

- **Description:** ____
- **Action Steps:**
  - Step 1: ____
  - Step 2: ____
  - Step 3: ____
- **Resources Needed:** ________________________________
- **Potential Challenges:** ____________________________
- **Success Metrics:** _________________________________
- **Deadline:** _______________________________________

Goal 2:

- **Description:** ____
- **Action Steps:**
  - Step 1: ____
  - Step 2: ____
  - Step 3: ____
- **Resources Needed:** ________________________________
- **Potential Challenges:** ____________________________
- **Success Metrics:** _________________________________
- **Deadline:** _______________________________________

Goal 3:

- **Description:** ______________________________________
- **Action Steps:**
  - Step 1: ______________________________________
  - Step 2: ______________________________________
  - Step 3: ______________________________________
- **Resources Needed:** _________________________________
- **Potential Challenges:** _____________________________
- **Success Metrics:** __________________________________
- **Deadline:** ____

Key Performance Indicators (KPIs)

1. **KPI 1:**
    - **Description:** ____
    - **Measurement Method:** ______________________________
    - **Target Value:** _______________________________________
2. **KPI 2:**
    - **Description:** ____
    - **Measurement Method:** ______________________________
    - **Target Value:** _______________________________________
3. **KPI 3:**
    - **Description:** ____
    - **Measurement Method:** ______________________________
    - **Target Value:** _______________________________________

Risk Management

1. **Risk 1:**
   - **Description:** ______________________________
   - **Mitigation Strategy:** ______________________________
   - **Impact Level:** ______________________________
   - **Likelihood:** ______________________________
2. **Risk 2:**
   - **Description:** ______________________________
   - **Mitigation Strategy:** ______________________________
   - **Impact Level:** ______________________________
   - **Likelihood:** ______________________________
3. **Risk 3:**
   - **Description:** ______________________________
   - **Mitigation Strategy:** ______________________________
   - **Impact Level:** ______________________________
   - **Likelihood:** ______________________________

Stakeholder Engagement

1. **Stakeholder 1:**
   - **Role:** ____
   - **Interest Level:** _____________________________________
   - **Communication Strategy:** ____________________________
2. **Stakeholder 2:**
   - **Role:** ____
   - **Interest Level:** _____________________________________
   - **Communication Strategy:** ____________________________
3. **Stakeholder 3:**
   - **Role:** ____
   - **Interest Level:** _____________________________________
   - **Communication Strategy:** ____________________________

Timeline

Task	Start Date	End Date	Responsible Person	Status
Task 1	___________	___________	____________________	_____________
Task 2	___________	___________	____________________	_____________
Task 3	___________	___________	____________________	_____________
Task 4	___________	___________	____________________	_____________

Budget

- **Total Budget:** ____
- **Cost Breakdown:**
    - **Personnel:** ____
    - **Technology:** ____
    - **Training:** ____
    - **Other Expenses:** _______________________________________

Review and Adjustments

- **Review Date 1:** ____
- **Adjustments Made:** ____
- **Review Date 2:** ____
- **Adjustments Made:** ____
- **Review Date 3:** ____
- **Adjustments Made:** ____

Set Budget for AI Project

Recreate this spreadsheet using the formulas below

Initial Investment:	$100,000.00
Operational Costs (Yearly):	$20,000.00
Revenue Impact (Yearly):	$200,000.00
Cost Savings (Yearly):	$50,000.00
COMPONENTS OF THE ROI FORMULA	
ROI=(Revenue Impact +Cost Savings)-(Initial Investment+Operational Costs)Initial Investment+Operational Costs	$130,000.00
Initial Investment: This is the upfront cost required to start the AI project. It includes expenses related to data gathering, software, hardware and team staffing	
Operational Costs: These are the ongoing costs incurred to maintain support and train the AI system on a yearly basis.	
Revenue Impact: This represents the additional revenue generated annually as a result of implementing the AI System. This could come from increased sales, improved customer retention or new revenue streams enabled by AI	
Cost Savings: These are the savings achieved annually through increased efficiency and accuracy due to the AI implementation. This might include reduced labor costs, lower error rates or more efficient resource usage.	
In order to obtain the numbers for inputting into our above formula we have to compute a few individual formulas first	
Calculate Total Benefits:	
Total Benefits=Revenue Impact+Annual Cost Savings	$250,000.00
Calculate Total Costs:	
Total Costs=Initial Investment+Operational Costs	$120,000.00

Calculate Net Benefits:	
Net Benefits=Total Benefits-Total Costs	$130,000.00
Calculate Return on Investment:	
ROI=Net Benefit/Total Costs	1.08
Calculate Return on Investment as a %:	
ROI%=ROIx100/100	108%
This means the AI project returns ____% of the investment annually which indicates whether the AI Project investment is profitable or not.	

Assemble Team for AI Project

Team Name	Person	Role

Create a Feasibility Study for each AI Project

This study aims to evaluate the feasibility of implementing an AI project within your business. It will analyze key factors to determine if the project is practical, viable, and aligns with your business goals.

I. Project Definition

- **Project Title:** Briefly describe your AI project in a clear and concise title.
- **Project Description:** In a few sentences, explain the overall purpose and functionality of your AI project.
- **Business Need:** Identify the specific business need your AI project aims to address. What problem will it solve or what opportunity will it capture?

II. Market Analysis

- **Target Market:** Identify your target market for the AI solution. Who will benefit from this project?
- **Competitive Landscape:** Research existing AI solutions in the market addressing similar problems. Are there any direct competitors? What are their strengths and weaknesses compared to your proposed solution?

III. Technical Feasibility

- **Data Availability:** Assess the data required for training and running your AI model. Do you have access to sufficient, high-quality data? If not, how will you acquire it?
- **Technical Expertise:** Evaluate your current technical resources. Do you have the in-house expertise to develop and maintain the AI project? If not, consider outsourcing options or acquiring the necessary skills.
- **AI Technology Selection:** Explore different AI technologies (Machine Learning, Deep Learning, etc.) suitable for your project goals. What are the strengths and limitations of each approach?

IV. Financial Feasibility

- **Cost Analysis:** Estimate the costs associated with your AI project. This includes hardware, software, data acquisition, development expenses, and potential ongoing maintenance costs.
- **Return on Investment (ROI):** Project the potential return on investment for your AI project. How will this project improve efficiency, increase revenue, or reduce costs? Develop a financial model to estimate the ROI timeframe.
- **Funding Options:** Identify potential funding sources for your AI project. Consider internal budgets, grants, or potential investors.

V. Risk Assessment

- **Identify Risks:** Outline potential risks associated with your AI project. This could include technical challenges, data security issues, ethical considerations, or potential project delays.
- **Mitigation Strategies:** Develop strategies to mitigate identified risks. How can you minimize the chances of these risks occurring and establish contingency plans if they do?

VI. Project Timeline

- **Develop a project timeline:** Create a realistic schedule outlining key milestones for the development and implementation of your AI project.

VII. Conclusion

- Based on the findings of this feasibility study, summarize your overall assessment. Is the AI project feasible for your business? If so, outline next steps for project development and implementation.
- If the project is not feasible at this time, explain the reasoning and suggest potential adjustments to the project scope or approach that could improve its feasibility.

Additional Considerations

- **Ethical Implications:** Consider the potential ethical implications of your AI project. Will it have any bias or privacy concerns that need to be addressed?
- **Regulatory Compliance:** Research any regulatory requirements that might apply to your AI project.

By conducting a thorough feasibility study, you can make informed decisions about the viability of your business AI project and increase the chances of its success.

Select AI Use Case Studies

Choose specific use cases that align with business objectives.

Use Case	Business Objective

Gather Relevant Data from Various Sources

Types of Data Collected: business records, interviews with key stakeholders, technology audits.

- **Customer Data:**____
- **Sales Data:**____
- **Operational Data:**____
- **Financial Data:**____

Data Collection Methods:

- **Manual Entry:**____
- **Automated Systems:**____

Data Storage and Management:

- **On-premises:**____
- **Cloud-based:**____
- **Hybrid:**____

Data Quality (Accuracy, Completeness, etc.): Use the data to assess the feasibility of the AI project, considering factors like current technology usage, business goals, data availability, and financial resources.

____

____

Data Quality through Cleaning and Preprocessing

Techniques to Ensure Data Quality, Accuracy, and Completeness

1. **Data Validation Rules:** Establish rules to ensure data accuracy during data entry.
2. **Regular Audits:** Conduct regular audits and reviews of data to identify and correct errors.
3. **Automated Workflows:** Use automated workflows to streamline data entry and reduce human errors.
4. **Data Standardization:** Standardize data formats and values to ensure consistency.
5. **Employee Training:** Train employees on best practices for data entry and management.

Clean and Preprocess Data

Ensuring the quality, accuracy, and completeness of business data is crucial for making informed decisions, especially when evaluating the feasibility of an AI project. Here are several tools and techniques that can be used to achieve this:

Data Quality Tools

1. **Data Profiling Tools:**
   - **IBM InfoSphere QualityStage:** Profiles, cleanses, and standardizes data.
   - Talend Data Preparation: Allows users to discover, cleanse, and format data.
   - Informatica Data Quality: Provides profiling, cleansing, and matching functions.
2. **Data Cleaning Tools:**
   - **OpenRefine:** An open-source tool for cleaning messy data.
   - Trifacta Wrangler: A data wrangling tool for cleaning and transforming raw data into structured formats.
   - **Data Ladder:** Helps in data matching, deduplication, and cleansing.
3. **Data Validation Tools:**
   - **Ataccama:** Provides automated data quality monitoring and validation.
   - **DataCleaner:** An open-source data quality analysis and data profiling tool.
   - **WinPure Clean & Match:** Validates and cleanses data by finding and removing duplicates.

Data Accuracy Tools

1. **Data Auditing Tools:**
   - **Apache Griffin:** An open-source data quality solution for big data environments.
   - **Datawatch Monarch:** Extracts and converts structured data from various sources and formats.
   - **Alteryx:** Combines data preparation, data blending, and analytics.
2. **Data Matching Tools:**
   - **Melissa Data:** Offers tools for data verification, address validation, and deduplication.
   - **MatchIt:** Provides data matching and deduplication services.

Data Completeness Tools

1. **Master Data Management (MDM) Tools:**
   - **SAP Master Data Governance:** Ensures data completeness and consistency across the organization.
   - **Oracle MDM:** Provides a comprehensive solution for managing master data.
   - **Stibo Systems:** Offers MDM solutions for various industries to ensure complete and accurate data.
2. **Data Integration Tools:**
   - **Microsoft SQL Server Integration Services (SSIS):** Integrates data from different sources and ensures completeness.

- **Apache Nifi:** An open-source tool for automating data flow between systems.
- **Talend Data Integration:** Integrates, transforms, and manages data across various platforms.

Data Governance Tools

1. **Collibra:** Provides data governance solutions to manage data assets, ensure compliance, and maintain data quality.
2. **Informatica Axon:** Helps in creating a collaborative data governance framework.
3. **Alation:** Combines data cataloging, data governance, and data stewardship.

Data Monitoring and Reporting Tools

1. **Tableau:** Visualizes data quality issues and provides insights into data completeness and accuracy.
2. **Power BI:** Monitors data quality and generates reports to ensure data accuracy and completeness.
3. **Qlik Sense:** Offers data visualization and reporting to identify and rectify data quality issues.

Develop and Train AI Models for Improved Accuracy and Performance

Developing and training AI models to improve the accuracy and performance of your data involves several key steps, from data preparation to model evaluation and deployment. Here's a structured approach:

Step 1: Data Collection and Preparation

1. **Data Collection:**
   - Gather data from multiple sources to ensure diversity and comprehensiveness.
   - Use APIs, web scraping, or data integration tools to aggregate data.
2. **Data Cleaning:**
   - Handle missing values using techniques like imputation or deletion.
   - Remove duplicates and irrelevant data.
   - Correct errors and standardize formats.
3. **Data Transformation:**
   - Normalize or standardize numerical data.
   - Encode categorical data using techniques like one-hot encoding.
   - Create new features from existing data (feature engineering).
4. **Data Splitting:**
   - Split data into training, validation, and test sets (commonly 70% training, 15% validation, 15% testing).

Step 2: Model Selection

1. **Choose Appropriate Algorithms:**
   - Based on the problem type (classification, regression, clustering, etc.), choose algorithms like decision trees, random forests, support vector machines (SVM), neural networks, etc.
2. **Baseline Model:**
   - Start with a simple baseline model to set a performance benchmark.

Step 3: Model Training

1. **Training:**
   - Train the model on the training dataset.
   - Use cross-validation to ensure the model generalizes well to unseen data.
2. **Hyperparameter Tuning:**
   - Use techniques like grid search or random search to find the best hyperparameters.
   - Consider using more advanced methods like Bayesian optimization.

Step 4: Model Evaluation

1. **Performance Metrics:**

   - Evaluate models using appropriate metrics such as accuracy, precision, recall, F1 score for classification, or mean squared error (MSE) for regression.
   - Use confusion matrices, ROC curves, and AUC scores for detailed analysis.
2. **Validation:**
   - Validate the model on the validation dataset to fine-tune and adjust.
3. **Test:**
   - Test the final model on the test dataset to evaluate its performance on unseen data.

Step 5: Model Improvement

1. **Feature Engineering:**
   - Create new features or select more relevant ones to improve model performance.
2. **Ensemble Methods:**
   - Combine multiple models using techniques like bagging, boosting, or stacking to improve accuracy.
3. **Regularization:**
   - Apply regularization techniques like L1 or L2 regularization to prevent overfitting.
4. **Data Augmentation:**
   - Use data augmentation techniques to artificially increase the size of the training dataset.

Step 6: Model Deployment and Monitoring

1. **Deployment:**
   - Deploy the model into a production environment using platforms like TensorFlow Serving, AWS SageMaker, or Docker.
2. **Monitoring:**
   - Continuously monitor the model's performance using metrics and logs.
   - Set up alerts for significant performance drops.
3. **Retraining:**
   - Periodically retrain the model with new data to keep it updated and accurate.

Step 7: Tools and Technologies

1. **Programming Languages:**
   - Use Python or R for data science and machine learning tasks.
2. **Libraries and Frameworks:**
   - **Scikit-learn:** For traditional machine learning algorithms.
   - **TensorFlow/Keras:** For building and training neural networks.
   - **PyTorch:** For deep learning and complex models.
   - **Pandas & NumPy:** For data manipulation and numerical operations.
3. **Model Management Tools:**
   - **MLflow:** For managing the ML lifecycle, including experimentation, reproducibility, and deployment.

- **DVC (Data Version Control):** For versioning datasets and machine learning models.
- **Kubeflow:** For deploying machine learning workflows on Kubernetes.

Implementation Checklists

Planning Phase

1. Define Objectives:

☐ Clearly outline the goals of your AI Project
☐ Ensure alignment with overall business strategy
☐ Identify specific problems or opportunities AI will address

2. Set Budget:

☐ Establish a budget for the project
☐ Include costs for data collection, tools, development, deployment and maintenance
☐ Account for potential contingency funds

3. Assemble Team:

☐ Identify key team members and define their roles
☐ Include data scientists, AI Engineers, project managers and business analysts
☐ Ensure cross-functional collaboration

4. Feasibility Study:

☐ Conduct a study to evaluate the feasibility of the AI project
☐ Assess technical, financial, and operational feasibility
☐ Identify potential risks and mitigation strategies

5. Select AI Use Cases:

☐ Choose specific use cases that align with business objectives
☐ Prioritize use cases based on impact and feasibility
☐ Develop a detailed plan for each use case

Development Phase

1. Data Collection:

- ☐ Gather relevant data from various sources
- ☐ Ensure data quality through cleaning and preprocessing
- ☐ Address any data gaps and inconsistencies

2. Select Tools and Platforms:

- ☐ Choose the AI Tools and platforms to be used
- ☐ Consider scalability, ease of use and compatibility with existing systems
- ☐ Evaluate open source vs. commercial options

3. Model Development

- ☐ Develop and train Ai Models
- ☐ Use appropriate algorithms and techniques
- ☐ Iterate on model development to improve accuracy and performance

4. Pilot Testing

- ☐ Conduct initial tests of the AI models
- ☐ Validate models with real-world data
- ☐ Gather feedback and make necessary adjustments

Deployment Phase

1. Integration:

- ☐ Integrate AI models into existing systems
- ☐ Ensure smooth interoperability with current workflows
- ☐ Address any integration challenges promptly

2. Scalability:

- ☐ Ensure systems can scale to meet increased demand
- ☐ Plan for additional computational resources if needed
- ☐ Optimize models for performance and efficiency

3. Monitor and Evaluation:

- ☐ Continuously monitor AI models and evaluate performance
- ☐ Set up automated alerts for any performance issues
- ☐ Regularly review metrics and adjust models as necessary

4. User Training:

- ☐ Train end-users on how to interact with and utilize the AI system
- ☐ Provide comprehensive training materials and sessions
- ☐ Encourage feedback and address any user concerns

Post-Deployment Phase

1. Feedback Loop:

- ☐ Establish a feedback loop for continuous improvement
- ☐ Regularly collect feedback from users and stakeholders
- ☐ Use feedback to refine and enhance AI models

2. Performance Metrics:

- ☐ Regularly review performance metrics
- ☐ Ensure models meet predefined success criteria
- ☐ Adjust metrics as business needs evolve

3. Periodic Audits:

- ☐ Conduct periodic audits to ensure compliance and performance
- ☐ Evaluate the ethical and legal implications of AI use
- ☐ Document findings and implement recommendations

Detailed Case Studies

Implementing AI in Retail for Inventory Management

Background: In the highly competitive world of retail, maintaining optimal inventory levels is crucial for ensuring customer satisfaction and maximizing sales. A leading clothing retailer, FashionPulse, was facing significant challenges with inventory management. Frequent stockouts led to lost sales opportunities, dissatisfied customers, and inefficient use of storage space. Recognizing the need for a transformative solution, FashionPulse decided to implement AI-powered predictive analytics to overhaul its inventory management system.

Objective: The primary objective was to leverage AI to improve inventory management through accurate demand forecasting. By doing so, FashionPulse aimed to reduce stockouts by 30% and increase overall sales by 15%.

Implementation:

1. **Data Collection and Preparation:** FashionPulse began by gathering extensive historical sales data, which included information on past sales, seasonal trends, customer preferences, and promotional activities. The company also integrated external data sources such as weather forecasts, regional events, and economic indicators to enhance the accuracy of the predictions.
2. **AI Model Development:** The retailer collaborated with a team of data scientists to develop a predictive analytics model. Using machine learning algorithms, the team built a model capable of analyzing the collected data and generating accurate demand forecasts. The initial model underwent rigorous testing and validation to ensure its reliability.
3. **Iterative Model Improvement:** Understanding that AI models require continuous refinement, FashionPulse adopted an iterative approach. The model was regularly updated with new data, and its performance was closely monitored. Feedback from inventory managers and store staff was used to make adjustments and improve the model's accuracy over time.
4. **Integration with Inventory Management System:** The AI-powered predictive analytics model was integrated with FashionPulse's existing inventory management system. This integration enabled automated stock replenishment based on real-time demand forecasts. The system provided store managers with actionable insights, allowing them to make informed decisions about inventory levels and product placements.

Outcome: The implementation of AI-powered predictive analytics brought significant positive changes to FashionPulse's inventory management:

- **Reduction in Stockouts:** The AI system's accurate demand forecasts enabled the retailer to maintain optimal inventory levels, reducing stockouts by 30%. This improvement meant that popular items were almost always available, enhancing customer satisfaction and loyalty.
- **Increase in Sales:** With fewer stockouts and better inventory management, FashionPulse saw a 15% increase in sales. Customers could find the products they wanted, leading to more completed purchases and higher revenue.

- **Operational Efficiency:** The automated stock replenishment system streamlined operations, reducing the manual workload for inventory managers. This efficiency allowed staff to focus on other important tasks, further improving the overall productivity of the business.

Key Lessons:

1. **Importance of High-Quality Data:** The success of the AI implementation hinged on the availability of high-quality data. Accurate and comprehensive data inputs were essential for generating reliable demand forecasts.
2. **Iterative Model Improvement:** Continuous improvement of the AI model was crucial. Regular updates and refinements based on new data and feedback ensured that the system remained accurate and relevant in a dynamic retail environment.
3. **Collaboration and Integration:** Successful implementation required collaboration between data scientists, IT professionals, and retail managers. Integrating the AI model with existing systems was key to seamless operations and maximizing the benefits of predictive analytics.

Conclusion: FashionPulse's journey of integrating AI into their inventory management system highlights the transformative potential of predictive analytics in retail. By leveraging AI, the retailer not only resolved its inventory challenges but also achieved significant business growth. The case underscores the necessity of high-quality data, continuous model improvement, and effective integration for successful AI implementation. As a result, FashionPulse set a new benchmark in the retail industry for inventory management, positioning itself for sustained success in the competitive market.

Implementing AI in Healthcare for Patient Care Optimization

Background: In the fast-paced and ever-evolving field of healthcare, patient care optimization is critical for improving outcomes and ensuring efficient use of resources. MediCarePlus, a leading healthcare provider, was facing challenges in managing patient care, leading to delayed treatments and suboptimal resource allocation. To address these issues, MediCarePlus decided to leverage AI-powered predictive analytics to enhance patient care and optimize resource utilization.

Objective: The primary objective was to utilize AI to improve patient care through accurate patient outcome predictions and efficient resource allocation. MediCarePlus aimed to reduce patient wait times by 25% and improve patient outcomes by 20%.

Implementation:

1. **Data Collection and Preparation:** MediCarePlus began by collecting a comprehensive dataset that included patient demographics, medical histories, treatment plans, and outcomes. Additionally, data on hospital resources, such as staff availability, equipment usage, and bed occupancy rates, were gathered to provide a complete picture of the healthcare environment.
2. **AI Model Development:** The healthcare provider collaborated with data scientists to develop predictive analytics models. These models utilized machine learning algorithms to analyze the collected data and generate accurate predictions about patient outcomes and resource needs. The initial models underwent extensive testing and validation to ensure their reliability and accuracy.
3. **Iterative Model Improvement:** Recognizing the importance of continuous improvement, MediCarePlus adopted an iterative approach to model development. The AI models were regularly updated with new data and feedback from healthcare professionals. This iterative process helped refine the models and improve their predictive capabilities over time.
4. **Integration with Patient Care Systems:** The AI-powered predictive analytics models were integrated with MediCarePlus's existing patient care management systems. This integration allowed for real-time monitoring of patient status and resource utilization. The system provided healthcare professionals with actionable insights, enabling them to make informed decisions about patient care and resource allocation.

Outcome: The implementation of AI-powered predictive analytics resulted in significant improvements in patient care and resource management at MediCarePlus:

- **Reduction in Patient Wait Times:** The AI system's accurate predictions allowed for better scheduling and resource allocation, leading to a 25% reduction in patient wait times. This improvement enhanced patient satisfaction and reduced the stress associated with long waits.
- **Improvement in Patient Outcomes:** By providing healthcare professionals with accurate predictions and actionable insights, the AI system helped improve patient

outcomes by 20%. Early identification of high-risk patients and timely interventions contributed to better recovery rates and overall patient health.
- **Operational Efficiency:** The automated system streamlined operations and reduced the manual workload for healthcare staff. Efficient resource allocation ensured that medical equipment and personnel were utilized effectively, leading to improved operational efficiency.

Key Lessons:

1. **Importance of High-Quality Data:** The success of the AI implementation was heavily dependent on the quality of the data collected. Comprehensive and accurate data inputs were crucial for generating reliable predictions and insights.
2. **Iterative Model Improvement:** Continuous refinement of the AI models was essential for maintaining accuracy and relevance. Regular updates based on new data and feedback from healthcare professionals ensured that the models adapted to changing conditions and needs.
3. **Collaboration and Integration:** Successful implementation required close collaboration between data scientists, IT professionals, and healthcare providers. Integrating the AI models with existing systems and workflows was key to achieving seamless operations and maximizing the benefits of predictive analytics.

Conclusion: MediCarePlus's journey of integrating AI into their patient care management system demonstrates the transformative potential of predictive analytics in healthcare. By leveraging AI, the healthcare provider not only addressed its challenges but also achieved significant improvements in patient care and operational efficiency. The case underscores the necessity of high-quality data, continuous model improvement, and effective integration for successful AI implementation. As a result, MediCarePlus set a new standard in the healthcare industry for patient care optimization, positioning itself as a leader in delivering high-quality healthcare services.

Implementing AI in Manufacturing for Predictive Maintenance

Background: In the highly competitive manufacturing industry, equipment downtime can lead to significant losses in productivity and revenue. PrecisionParts Inc., a leading manufacturer of automotive components, was experiencing frequent equipment failures that disrupted production schedules and increased operational costs. To tackle this challenge, PrecisionParts Inc. decided to implement AI-powered predictive maintenance to enhance equipment reliability and optimize maintenance schedules.

Objective: The primary objective was to leverage AI to predict equipment failures before they occurred, allowing for timely maintenance and reducing unplanned downtime. PrecisionParts Inc. aimed to decrease equipment downtime by 40% and lower maintenance costs by 25%.

Implementation:

1. **Data Collection and Preparation:** PrecisionParts Inc. began by collecting extensive data from their manufacturing equipment, including sensor data, operational logs, maintenance records, and failure reports. This data was crucial for training the predictive maintenance models.
2. **AI Model Development:** The company collaborated with data scientists to develop machine learning models capable of analyzing the collected data and predicting equipment failures. The models were designed to identify patterns and anomalies that indicated potential issues. Rigorous testing and validation ensured the models' accuracy and reliability.
3. **Iterative Model Improvement:** Understanding the need for continuous improvement, PrecisionParts Inc. adopted an iterative approach to model development. The predictive maintenance models were regularly updated with new data and feedback from maintenance engineers, refining their accuracy over time.
4. **Integration with Maintenance Management Systems:** The AI-powered predictive maintenance models were integrated with the company's existing maintenance management systems. This integration allowed for real-time monitoring of equipment health and automated maintenance scheduling based on the AI predictions. Maintenance teams received alerts and actionable insights, enabling proactive maintenance actions.

Outcome: The implementation of AI-powered predictive maintenance led to significant improvements in equipment reliability and operational efficiency at PrecisionParts Inc.:

- **Reduction in Equipment Downtime:** The AI system's accurate predictions enabled timely maintenance, resulting in a 40% reduction in equipment downtime. This improvement ensured smoother production schedules and higher productivity.
- **Decrease in Maintenance Costs:** By predicting failures and optimizing maintenance schedules, the company reduced unnecessary maintenance activities and minimized emergency repairs, leading to a 25% decrease in maintenance costs.
- **Enhanced Equipment Reliability:** Proactive maintenance based on AI predictions improved the overall reliability and lifespan of manufacturing equipment, reducing the frequency of unexpected breakdowns.

Key Lessons:

1. **Importance of High-Quality Data:** The success of the AI implementation was heavily dependent on the quality and comprehensiveness of the data collected. Accurate and detailed data inputs were essential for reliable predictions.
2. **Iterative Model Improvement:** Continuous refinement of the predictive maintenance models was crucial for maintaining their accuracy and effectiveness. Regular updates based on new data and feedback from maintenance teams ensured the models adapted to changing conditions and needs.
3. **Collaboration and Integration:** Successful implementation required close collaboration between data scientists, IT professionals, and maintenance engineers. Integrating the AI models with existing systems and workflows was key to achieving seamless operations and maximizing the benefits of predictive maintenance.

Conclusion: PrecisionParts Inc.'s journey of integrating AI into their maintenance management system highlights the transformative potential of predictive maintenance in the manufacturing industry. By leveraging AI, the company not only addressed its equipment reliability challenges but also achieved significant improvements in operational efficiency and cost savings. The case underscores the necessity of high-quality data, continuous model improvement, and effective integration for successful AI implementation. As a result, PrecisionParts Inc. set a new standard in the manufacturing industry for predictive maintenance, positioning itself as a leader in innovative and efficient manufacturing practices.

Implementing AI in Finance for Fraud Detection

Background: In the finance industry, fraud detection is critical to protecting assets and maintaining trust with customers. SecureBank, a leading financial institution, was facing increasing challenges with detecting and preventing fraudulent transactions. Traditional rule-based systems were insufficient in keeping up with the sophisticated methods used by fraudsters. To enhance their fraud detection capabilities, SecureBank decided to implement AI-powered predictive analytics.

Objective: The primary objective was to leverage AI to improve fraud detection accuracy and reduce false positives. SecureBank aimed to increase the detection rate of fraudulent transactions by 50% and decrease false positives by 30%.

Implementation:

1. **Data Collection and Preparation:** SecureBank began by collecting a vast amount of data, including transaction histories, customer profiles, device information, and known fraud patterns. The data was meticulously cleaned and labeled to ensure high-quality inputs for training the AI models.
2. **AI Model Development:** The bank collaborated with data scientists to develop machine learning models designed to detect fraudulent activities. These models analyzed patterns and anomalies in transaction data to identify potential fraud. The initial models underwent rigorous testing and validation to ensure their accuracy and reliability.
3. **Iterative Model Improvement:** Recognizing the dynamic nature of fraud, SecureBank adopted an iterative approach to model development. The AI models were regularly updated with new data and feedback from fraud analysts, allowing for continuous improvement in detection capabilities.
4. **Integration with Fraud Detection Systems:** The AI-powered models were integrated with SecureBank's existing fraud detection systems. This integration enabled real-time analysis of transactions and immediate flagging of suspicious activities. The system provided fraud analysts with actionable insights, allowing for quick and informed decision-making.

Outcome: The implementation of AI-powered fraud detection resulted in significant improvements in SecureBank's ability to detect and prevent fraudulent transactions:

- **Increase in Fraud Detection Rate:** The AI system's sophisticated algorithms improved the detection rate of fraudulent transactions by 50%. This enhancement helped SecureBank stay ahead of evolving fraud tactics and protect its customers more effectively.
- **Reduction in False Positives:** By accurately distinguishing between legitimate and fraudulent transactions, the AI system reduced false positives by 30%. This reduction minimized disruptions for customers and improved the efficiency of fraud investigations.
- **Enhanced Operational Efficiency:** The automated system streamlined the fraud detection process, reducing the manual workload for fraud analysts. This efficiency allowed analysts to focus on more complex cases and improve overall response times.

Key Lessons:

1. **Importance of High-Quality Data:** The success of the AI implementation was heavily dependent on the quality and comprehensiveness of the data collected. Accurate and detailed data inputs were essential for reliable fraud detection.
2. **Iterative Model Improvement:** Continuous refinement of the AI models was crucial for maintaining their accuracy and effectiveness. Regular updates based on new data and feedback from fraud analysts ensured the models adapted to emerging fraud patterns.
3. **Collaboration and Integration:** Successful implementation required close collaboration between data scientists, IT professionals, and fraud analysts. Integrating the AI models with existing systems and workflows was key to achieving seamless operations and maximizing the benefits of predictive analytics.

Conclusion: SecureBank's journey of integrating AI into their fraud detection system highlights the transformative potential of predictive analytics in the finance industry. By leveraging AI, the bank not only enhanced its fraud detection capabilities but also achieved significant improvements in operational efficiency and customer satisfaction. The case underscores the necessity of high-quality data, continuous model improvement, and effective integration for successful AI implementation. As a result, SecureBank set a new standard in the finance industry for fraud detection, positioning itself as a leader in innovative and secure financial services.

Implementing AI in Transportation for Fleet Management

Background: In the transportation industry, efficient fleet management is crucial for reducing costs and improving service quality. SpeedyTransit, a leading logistics and transportation company, was facing challenges in managing its fleet of vehicles. Frequent breakdowns, inefficient routes, and high fuel costs were hampering operational efficiency and increasing expenses. To address these issues, SpeedyTransit decided to implement AI-powered predictive analytics to optimize fleet management.

Objective: The primary objective was to leverage AI to improve fleet maintenance and optimize routing. SpeedyTransit aimed to reduce vehicle breakdowns by 40% and decrease fuel consumption by 20%.

Implementation:

1. **Data Collection and Preparation:** SpeedyTransit began by collecting extensive data from their fleet, including vehicle maintenance records, GPS data, fuel consumption logs, and driver behavior data. This data was essential for training the predictive analytics models.
2. **AI Model Development:** The company collaborated with data scientists to develop machine learning models capable of analyzing the collected data to predict vehicle maintenance needs and optimize routing. The models were designed to identify patterns and anomalies that indicated potential issues and inefficiencies. Rigorous testing and validation ensured the models' accuracy and reliability.
3. **Iterative Model Improvement:** Recognizing the importance of continuous improvement, SpeedyTransit adopted an iterative approach to model development. The predictive maintenance and routing optimization models were regularly updated with new data and feedback from fleet managers, refining their accuracy over time.
4. **Integration with Fleet Management Systems:** The AI-powered models were integrated with the company's existing fleet management systems. This integration allowed for real-time monitoring of vehicle health and optimized route planning. The system provided fleet managers with actionable insights, enabling proactive maintenance and efficient route scheduling.

Outcome: The implementation of AI-powered predictive analytics led to significant improvements in fleet management at SpeedyTransit:

- **Reduction in Vehicle Breakdowns:** The AI system's accurate predictions enabled timely maintenance, resulting in a 40% reduction in vehicle breakdowns. This improvement ensured more reliable service and reduced downtime.
- **Decrease in Fuel Consumption:** By optimizing routes and improving driving behavior through real-time feedback, the company decreased fuel consumption by 20%. This reduction led to significant cost savings and a smaller environmental footprint.
- **Enhanced Operational Efficiency:** The automated system streamlined operations and reduced the manual workload for fleet managers. Efficient route planning ensured timely deliveries and improved overall productivity.

Key Lessons:

1. **Importance of High-Quality Data:** The success of the AI implementation was heavily dependent on the quality and comprehensiveness of the data collected. Accurate and detailed data inputs were essential for reliable predictions and optimizations.
2. **Iterative Model Improvement:** Continuous refinement of the predictive maintenance and routing optimization models was crucial for maintaining their accuracy and effectiveness. Regular updates based on new data and feedback from fleet managers ensured the models adapted to changing conditions and needs.
3. **Collaboration and Integration:** Successful implementation required close collaboration between data scientists, IT professionals, and fleet managers. Integrating the AI models with existing systems and workflows was key to achieving seamless operations and maximizing the benefits of predictive analytics.

Conclusion: SpeedyTransit's journey of integrating AI into their fleet management system highlights the transformative potential of predictive analytics in the transportation industry. By leveraging AI, the company not only addressed its fleet management challenges but also achieved significant improvements in operational efficiency and cost savings. The case underscores the necessity of high-quality data, continuous model improvement, and effective integration for successful AI implementation. As a result, SpeedyTransit set a new standard in the transportation industry for fleet management, positioning itself as a leader in innovative and efficient logistics services.

Implementing AI in Energy for Predictive Maintenance and Efficiency Optimization

Background: In the energy industry, maintaining equipment reliability and optimizing energy production are critical for ensuring operational efficiency and reducing costs. PowerGen, a leading energy company specializing in renewable energy sources, was facing challenges with equipment maintenance and energy production efficiency. Frequent equipment failures and suboptimal energy output were increasing operational costs and reducing profitability. To tackle these issues, PowerGen decided to implement AI-powered predictive analytics to enhance equipment maintenance and optimize energy production.

Objective: The primary objective was to leverage AI to predict equipment failures and optimize energy production. PowerGen aimed to decrease equipment downtime by 35% and increase energy production efficiency by 25%.

Implementation:

1. **Data Collection and Preparation:** PowerGen began by collecting extensive data from their energy production equipment, including sensor data, operational logs, maintenance records, and environmental conditions. This data was essential for training the predictive analytics models.
2. **AI Model Development:** The company collaborated with data scientists to develop machine learning models capable of analyzing the collected data to predict equipment maintenance needs and optimize energy production. The models were designed to identify patterns and anomalies that indicated potential issues and inefficiencies. Rigorous testing and validation ensured the models' accuracy and reliability.
3. **Iterative Model Improvement:** Recognizing the importance of continuous improvement, PowerGen adopted an iterative approach to model development. The predictive maintenance and energy optimization models were regularly updated with new data and feedback from engineers, refining their accuracy over time.
4. **Integration with Maintenance and Production Systems:** The AI-powered models were integrated with the company's existing maintenance and production management systems. This integration allowed for real-time monitoring of equipment health and energy production optimization. The system provided engineers with actionable insights, enabling proactive maintenance and efficient energy production scheduling.

Outcome: The implementation of AI-powered predictive analytics led to significant improvements in equipment reliability and energy production efficiency at PowerGen:

- **Reduction in Equipment Downtime:** The AI system's accurate predictions enabled timely maintenance, resulting in a 35% reduction in equipment downtime. This improvement ensured more reliable energy production and reduced operational disruptions.
- **Increase in Energy Production Efficiency:** By optimizing energy production processes and improving equipment performance, the company increased energy production

efficiency by 25%. This enhancement led to higher energy output and improved profitability.

- **Enhanced Operational Efficiency:** The automated system streamlined operations and reduced the manual workload for engineers. Efficient maintenance scheduling and energy production optimization ensured smooth operations and improved overall productivity.

Key Lessons:

1. **Importance of High-Quality Data:** The success of the AI implementation was heavily dependent on the quality and comprehensiveness of the data collected. Accurate and detailed data inputs were essential for reliable predictions and optimizations.
2. **Iterative Model Improvement:** Continuous refinement of the predictive maintenance and energy optimization models was crucial for maintaining their accuracy and effectiveness. Regular updates based on new data and feedback from engineers ensured the models adapted to changing conditions and needs.
3. **Collaboration and Integration:** Successful implementation required close collaboration between data scientists, IT professionals, and engineers. Integrating the AI models with existing systems and workflows was key to achieving seamless operations and maximizing the benefits of predictive analytics.

Conclusion: PowerGen's journey of integrating AI into their maintenance and production management systems highlights the transformative potential of predictive analytics in the energy industry. By leveraging AI, the company not only addressed its equipment maintenance and energy production challenges but also achieved significant improvements in operational efficiency and cost savings. The case underscores the necessity of high-quality data, continuous model improvement, and effective integration for successful AI implementation. As a result, PowerGen set a new standard in the energy industry for predictive maintenance and efficiency optimization, positioning itself as a leader in innovative and sustainable energy solutions.

Implementing AI in Telecommunications for Network Optimization

Background: In the telecommunications industry, ensuring network reliability and optimizing service delivery are crucial for maintaining customer satisfaction and operational efficiency. TeleComX, a leading telecommunications provider, was experiencing frequent network outages and suboptimal service quality, leading to increased customer complaints and operational costs. To address these challenges, TeleComX decided to implement AI-powered predictive analytics to enhance network performance and optimize resource allocation.

Objective: The primary objective was to leverage AI to predict network issues and optimize service delivery. TeleComX aimed to reduce network outages by 40% and improve overall service quality by 30%.

Implementation:

1. **Data Collection and Preparation:** TeleComX began by collecting extensive data from their network infrastructure, including performance metrics, usage patterns, maintenance records, and environmental conditions. This data was crucial for training the predictive analytics models.
2. **AI Model Development:** The company collaborated with data scientists to develop machine learning models capable of analyzing the collected data to predict network issues and optimize resource allocation. The models were designed to identify patterns and anomalies that indicated potential problems and inefficiencies. Rigorous testing and validation ensured the models' accuracy and reliability.
3. **Iterative Model Improvement:** Recognizing the importance of continuous improvement, TeleComX adopted an iterative approach to model development. The predictive maintenance and network optimization models were regularly updated with new data and feedback from network engineers, refining their accuracy over time.
4. **Integration with Network Management Systems:** The AI-powered models were integrated with the company's existing network management systems. This integration allowed for real-time monitoring of network performance and automated resource allocation based on AI predictions. The system provided network engineers with actionable insights, enabling proactive maintenance and efficient service delivery.

Outcome: The implementation of AI-powered predictive analytics led to significant improvements in network performance and operational efficiency at TeleComX:

- **Reduction in Network Outages:** The AI system's accurate predictions enabled timely maintenance and proactive issue resolution, resulting in a 40% reduction in network outages. This improvement ensured more reliable service and reduced customer complaints.
- **Improvement in Service Quality:** By optimizing resource allocation and improving network performance, the company improved overall service quality by 30%. This enhancement led to higher customer satisfaction and loyalty.
- **Enhanced Operational Efficiency:** The automated system streamlined operations and reduced the manual workload for network engineers. Efficient maintenance scheduling

and resource allocation ensured smooth network operations and improved overall productivity.

Key Lessons:

1. **Importance of High-Quality Data:** The success of the AI implementation was heavily dependent on the quality and comprehensiveness of the data collected. Accurate and detailed data inputs were essential for reliable predictions and optimizations.
2. **Iterative Model Improvement:** Continuous refinement of the predictive maintenance and network optimization models was crucial for maintaining their accuracy and effectiveness. Regular updates based on new data and feedback from network engineers ensured the models adapted to changing conditions and needs.
3. **Collaboration and Integration:** Successful implementation required close collaboration between data scientists, IT professionals, and network engineers. Integrating the AI models with existing systems and workflows was key to achieving seamless operations and maximizing the benefits of predictive analytics.

Conclusion: TeleComX's journey of integrating AI into their network management systems highlights the transformative potential of predictive analytics in the telecommunications industry. By leveraging AI, the company not only addressed its network performance challenges but also achieved significant improvements in operational efficiency and customer satisfaction. The case underscores the necessity of high-quality data, continuous model improvement, and effective integration for successful AI implementation. As a result, TeleComX set a new standard in the telecommunications industry for network optimization, positioning itself as a leader in innovative and reliable telecommunications services.

Implementing AI in Education for Personalized Learning

Background: In the education sector, providing personalized learning experiences is essential for improving student outcomes and engagement. EduLearn, a leading provider of online education platforms, was facing challenges with delivering tailored educational content to diverse student populations. Traditional one-size-fits-all approaches were insufficient in addressing the unique needs and learning paces of individual students. To tackle these issues, EduLearn decided to implement AI-powered personalized learning solutions.

Objective: The primary objective was to leverage AI to create personalized learning experiences that adapt to each student's needs. EduLearn aimed to increase student engagement by 35% and improve learning outcomes by 25%.

Implementation:

1. **Data Collection and Preparation:** EduLearn began by collecting extensive data on student interactions with the platform, including quiz scores, assignment submissions, time spent on different topics, and engagement metrics. This data was crucial for training the personalized learning models.
2. **AI Model Development:** The company collaborated with data scientists to develop machine learning models capable of analyzing the collected data to create personalized learning paths. The models were designed to identify patterns in student behavior and performance, recommending content that best suits each student's learning style. Rigorous testing and validation ensured the models' accuracy and reliability.
3. **Iterative Model Improvement:** Recognizing the importance of continuous improvement, EduLearn adopted an iterative approach to model development. The personalized learning models were regularly updated with new data and feedback from educators, refining their accuracy over time.
4. **Integration with Learning Management Systems:** The AI-powered models were integrated with the company's existing learning management systems. This integration allowed for real-time personalization of educational content based on AI predictions. The system provided educators with actionable insights, enabling them to support students more effectively.

Outcome: The implementation of AI-powered personalized learning solutions led to significant improvements in student engagement and learning outcomes at EduLearn:

- **Increase in Student Engagement:** The AI system's ability to tailor content to individual needs resulted in a 35% increase in student engagement. Students were more motivated and invested in their learning, leading to higher participation rates.
- **Improvement in Learning Outcomes:** By providing personalized learning paths that adapt to each student's pace and style, the company improved learning outcomes by 25%. Students achieved better grades and demonstrated a deeper understanding of the material.
- **Enhanced Educator Efficiency:** The automated system streamlined the process of personalizing content, reducing the manual workload for educators. This efficiency allowed educators to focus more on providing support and feedback to students.

Key Lessons:

1. **Importance of High-Quality Data:** The success of the AI implementation was heavily dependent on the quality and comprehensiveness of the data collected. Accurate and detailed data inputs were essential for reliable personalization.
2. **Iterative Model Improvement:** Continuous refinement of the personalized learning models was crucial for maintaining their accuracy and effectiveness. Regular updates based on new data and feedback from educators ensured the models adapted to changing student needs.
3. **Collaboration and Integration:** Successful implementation required close collaboration between data scientists, IT professionals, and educators. Integrating the AI models with existing systems and workflows was key to achieving seamless operations and maximizing the benefits of personalized learning.

Conclusion: EduLearn's journey of integrating AI into their learning management systems highlights the transformative potential of personalized learning in the education sector. By leveraging AI, the company not only addressed its challenges but also achieved significant improvements in student engagement and learning outcomes. The case underscores the necessity of high-quality data, continuous model improvement, and effective integration for successful AI implementation. As a result, EduLearn set a new standard in the education industry for personalized learning, positioning itself as a leader in innovative and effective educational solutions.

Implementing AI in Agriculture for Crop Yield Optimization

Background: In the agriculture industry, optimizing crop yields is crucial for ensuring food security and maximizing profits. AgroTech, a leading agricultural technology company, was facing challenges with managing crop health and predicting yields. Traditional farming methods were insufficient in addressing the complex variables that affect crop growth, such as weather patterns, soil conditions, and pest infestations. To tackle these issues, AgroTech decided to implement AI-powered predictive analytics to enhance crop yield optimization.

Objective: The primary objective was to leverage AI to predict crop yields and optimize farming practices. AgroTech aimed to increase crop yields by 30% and reduce resource usage (water, fertilizers, pesticides) by 20%.

Implementation:

1. **Data Collection and Preparation:** AgroTech began by collecting extensive data from their farms, including soil moisture levels, weather conditions, crop health imagery, and historical yield data. This data was crucial for training the predictive analytics models.
2. **AI Model Development:** The company collaborated with data scientists to develop machine learning models capable of analyzing the collected data to predict crop yields and optimize farming practices. The models were designed to identify patterns and correlations that influence crop growth. Rigorous testing and validation ensured the models' accuracy and reliability.
3. **Iterative Model Improvement:** Recognizing the importance of continuous improvement, AgroTech adopted an iterative approach to model development. The predictive models were regularly updated with new data and feedback from agronomists, refining their accuracy over time.
4. **Integration with Farm Management Systems:** The AI-powered models were integrated with the company's existing farm management systems. This integration allowed for real-time monitoring of crop health and resource usage, and automated adjustments based on AI predictions. The system provided farmers with actionable insights, enabling them to make informed decisions about planting, irrigation, fertilization, and pest control.

Outcome: The implementation of AI-powered predictive analytics led to significant improvements in crop yield and resource efficiency at AgroTech:

- **Increase in Crop Yields:** The AI system's accurate predictions and optimized farming practices resulted in a 30% increase in crop yields. This improvement ensured higher productivity and profitability for the farmers.
- **Reduction in Resource Usage:** By optimizing irrigation, fertilization, and pest control based on AI recommendations, the company reduced resource usage by 20%. This reduction led to cost savings and a smaller environmental footprint.
- **Enhanced Operational Efficiency:** The automated system streamlined operations and reduced the manual workload for farmers. Efficient resource management and real-time monitoring ensured smoother farming operations and improved overall productivity.

Key Lessons:

1. **Importance of High-Quality Data:** The success of the AI implementation was heavily dependent on the quality and comprehensiveness of the data collected. Accurate and detailed data inputs were essential for reliable predictions and optimizations.
2. **Iterative Model Improvement:** Continuous refinement of the predictive models was crucial for maintaining their accuracy and effectiveness. Regular updates based on new data and feedback from agronomists ensured the models adapted to changing conditions and needs.
3. **Collaboration and Integration:** Successful implementation required close collaboration between data scientists, IT professionals, and agronomists. Integrating the AI models with existing systems and workflows was key to achieving seamless operations and maximizing the benefits of predictive analytics.

Conclusion: AgroTech's journey of integrating AI into their farm management systems highlights the transformative potential of predictive analytics in the agriculture industry. By leveraging AI, the company not only addressed its challenges but also achieved significant improvements in crop yield and resource efficiency. The case underscores the necessity of high-quality data, continuous model improvement, and effective integration for successful AI implementation. As a result, AgroTech set a new standard in the agriculture industry for crop yield optimization, positioning itself as a leader in innovative and sustainable farming practices.

Implementing AI in Insurance for Risk Assessment and Fraud Detection

Background: In the insurance industry, accurate risk assessment and fraud detection are critical for maintaining profitability and customer trust. InsureTech, a leading insurance company, was facing challenges with evaluating risk and detecting fraudulent claims. Traditional methods were insufficient in addressing the complex variables involved, leading to higher claim payouts and increased operational costs. To tackle these issues, InsureTech decided to implement AI-powered predictive analytics to enhance risk assessment and fraud detection.

Objective: The primary objective was to leverage AI to improve risk assessment accuracy and detect fraudulent claims more effectively. InsureTech aimed to reduce fraudulent claim payouts by 40% and improve risk assessment accuracy by 25%.

Implementation:

1. **Data Collection and Preparation:** InsureTech began by collecting extensive data, including historical claims, customer profiles, policy details, and external data sources such as economic indicators and social media activity. This data was crucial for training the predictive analytics models.
2. **AI Model Development:** The company collaborated with data scientists to develop machine learning models capable of analyzing the collected data to predict risk levels and identify fraudulent claims. The models were designed to identify patterns and anomalies indicative of fraud and accurately assess risk. Rigorous testing and validation ensured the models' accuracy and reliability.
3. **Iterative Model Improvement:** Recognizing the importance of continuous improvement, InsureTech adopted an iterative approach to model development. The predictive models were regularly updated with new data and feedback from underwriters and fraud analysts, refining their accuracy over time.
4. **Integration with Insurance Management Systems:** The AI-powered models were integrated with the company's existing insurance management systems. This integration allowed for real-time analysis of claims and automated risk assessment based on AI predictions. The system provided underwriters and fraud analysts with actionable insights, enabling proactive measures to prevent fraud and accurately assess risk.

Outcome: The implementation of AI-powered predictive analytics led to significant improvements in risk assessment and fraud detection at InsureTech:

- **Reduction in Fraudulent Claim Payouts:** The AI system's accurate predictions and fraud detection capabilities resulted in a 40% reduction in fraudulent claim payouts. This improvement ensured better financial stability and reduced unnecessary costs.
- **Improvement in Risk Assessment Accuracy:** By leveraging AI for risk assessment, the company improved accuracy by 25%. This enhancement allowed InsureTech to price policies more accurately and manage risk more effectively.
- **Enhanced Operational Efficiency:** The automated system streamlined operations and reduced the manual workload for underwriters and fraud analysts. Efficient claim

processing and real-time risk assessment ensured smoother operations and improved overall productivity.

Key Lessons:

1. **Importance of High-Quality Data:** The success of the AI implementation was heavily dependent on the quality and comprehensiveness of the data collected. Accurate and detailed data inputs were essential for reliable predictions and fraud detection.
2. **Iterative Model Improvement:** Continuous refinement of the predictive models was crucial for maintaining their accuracy and effectiveness. Regular updates based on new data and feedback from underwriters and fraud analysts ensured the models adapted to changing conditions and needs.
3. **Collaboration and Integration:** Successful implementation required close collaboration between data scientists, IT professionals, and insurance experts. Integrating the AI models with existing systems and workflows was key to achieving seamless operations and maximizing the benefits of predictive analytics.

Conclusion: InsureTech's journey of integrating AI into their insurance management systems highlights the transformative potential of predictive analytics in the insurance industry. By leveraging AI, the company not only addressed its risk assessment and fraud detection challenges but also achieved significant improvements in operational efficiency and financial stability. The case underscores the necessity of high-quality data, continuous model improvement, and effective integration for successful AI implementation. As a result, InsureTech set a new standard in the insurance industry for risk assessment and fraud detection, positioning itself as a leader in innovative and reliable insurance services.

Implementing AI in Real Estate for Property Valuation and Market Analysis

Background: In the real estate industry, accurate property valuation and market analysis are crucial for making informed investment decisions and providing clients with reliable information. RealtyAnalytics, a leading real estate firm, was facing challenges with assessing property values and analyzing market trends accurately. Traditional methods were time-consuming and often imprecise, leading to missed opportunities and reduced client satisfaction. To address these issues, RealtyAnalytics decided to implement AI-powered predictive analytics to enhance property valuation and market analysis.

Objective: The primary objective was to leverage AI to improve the accuracy of property valuations and market analyses. RealtyAnalytics aimed to increase the accuracy of property valuations by 30% and reduce the time spent on market analysis by 50%.

Implementation:

1. **Data Collection and Preparation:** RealtyAnalytics began by collecting extensive data on properties, including historical sale prices, property features, location data, and market trends. Additional data sources included economic indicators, neighborhood statistics, and social factors. This data was crucial for training the predictive analytics models.
2. **AI Model Development:** The company collaborated with data scientists to develop machine learning models capable of analyzing the collected data to predict property values and analyze market trends. The models were designed to identify patterns and correlations that influence property prices. Rigorous testing and validation ensured the models' accuracy and reliability.
3. **Iterative Model Improvement:** Recognizing the importance of continuous improvement, RealtyAnalytics adopted an iterative approach to model development. The predictive models were regularly updated with new data and feedback from real estate agents, refining their accuracy over time.
4. **Integration with Real Estate Management Systems:** The AI-powered models were integrated with the company's existing real estate management systems. This integration allowed for real-time property valuation and market analysis based on AI predictions. The system provided real estate agents with actionable insights, enabling them to make informed decisions and provide better service to clients.

Outcome: The implementation of AI-powered predictive analytics led to significant improvements in property valuation and market analysis at RealtyAnalytics:

- **Increase in Property Valuation Accuracy:** The AI system's accurate predictions resulted in a 30% increase in property valuation accuracy. This improvement ensured more precise pricing and better investment decisions.
- **Reduction in Time Spent on Market Analysis:** By leveraging AI for market analysis, the company reduced the time spent on this task by 50%. This enhancement allowed real estate agents to focus more on client interactions and strategic planning.

- **Enhanced Operational Efficiency:** The automated system streamlined operations and reduced the manual workload for real estate agents. Efficient property valuation and market analysis ensured smoother operations and improved overall productivity.

Key Lessons:

1. **Importance of High-Quality Data:** The success of the AI implementation was heavily dependent on the quality and comprehensiveness of the data collected. Accurate and detailed data inputs were essential for reliable property valuations and market analyses.
2. **Iterative Model Improvement:** Continuous refinement of the predictive models was crucial for maintaining their accuracy and effectiveness. Regular updates based on new data and feedback from real estate agents ensured the models adapted to changing market conditions and client needs.
3. **Collaboration and Integration:** Successful implementation required close collaboration between data scientists, IT professionals, and real estate experts. Integrating the AI models with existing systems and workflows was key to achieving seamless operations and maximizing the benefits of predictive analytics.

Conclusion: RealtyAnalytics's journey of integrating AI into their real estate management systems highlights the transformative potential of predictive analytics in the real estate industry. By leveraging AI, the company not only addressed its property valuation and market analysis challenges but also achieved significant improvements in operational efficiency and client satisfaction. The case underscores the necessity of high-quality data, continuous model improvement, and effective integration for successful AI implementation. As a result, RealtyAnalytics set a new standard in the real estate industry for property valuation and market analysis, positioning itself as a leader in innovative and reliable real estate services.

Implementing AI in Automotive for Predictive Maintenance and Customer Experience Enhancement

Background: In the automotive industry, maintaining vehicle reliability and enhancing customer experience are crucial for ensuring customer satisfaction and loyalty. AutoTech, a leading automotive manufacturer, was facing challenges with vehicle maintenance and providing personalized customer services. Traditional methods were insufficient in predicting maintenance needs and personalizing customer interactions, leading to increased downtime and reduced customer satisfaction. To tackle these issues, AutoTech decided to implement AI-powered predictive analytics to enhance predictive maintenance and improve customer experience.

Objective: The primary objective was to leverage AI to predict vehicle maintenance needs and personalize customer services. AutoTech aimed to reduce vehicle downtime by 40% and increase customer satisfaction by 30%.

Implementation:

1. **Data Collection and Preparation:** AutoTech began by collecting extensive data from their vehicles, including sensor data, maintenance records, driving patterns, and customer feedback. This data was crucial for training the predictive analytics models.
2. **AI Model Development:** The company collaborated with data scientists to develop machine learning models capable of analyzing the collected data to predict maintenance needs and personalize customer interactions. The models were designed to identify patterns and anomalies that indicated potential issues and preferences. Rigorous testing and validation ensured the models' accuracy and reliability.
3. **Iterative Model Improvement:** Recognizing the importance of continuous improvement, AutoTech adopted an iterative approach to model development. The predictive models were regularly updated with new data and feedback from mechanics and customer service representatives, refining their accuracy over time.
4. **Integration with Automotive Management Systems:** The AI-powered models were integrated with the company's existing automotive management systems. This integration allowed for real-time monitoring of vehicle health and automated customer service recommendations based on AI predictions. The system provided mechanics and customer service representatives with actionable insights, enabling proactive maintenance and personalized customer interactions.

Outcome: The implementation of AI-powered predictive analytics led to significant improvements in vehicle maintenance and customer experience at AutoTech:

- **Reduction in Vehicle Downtime:** The AI system's accurate predictions enabled timely maintenance, resulting in a 40% reduction in vehicle downtime. This improvement ensured more reliable vehicles and reduced inconvenience for customers.
- **Increase in Customer Satisfaction:** By providing personalized services based on AI recommendations, the company increased customer satisfaction by 30%. Customers received tailored maintenance reminders and personalized service offers, enhancing their overall experience.

- **Enhanced Operational Efficiency:** The automated system streamlined operations and reduced the manual workload for mechanics and customer service representatives. Efficient maintenance scheduling and personalized customer interactions ensured smoother operations and improved overall productivity.

Key Lessons:

1. **Importance of High-Quality Data:** The success of the AI implementation was heavily dependent on the quality and comprehensiveness of the data collected. Accurate and detailed data inputs were essential for reliable predictions and personalized services.
2. **Iterative Model Improvement:** Continuous refinement of the predictive models was crucial for maintaining their accuracy and effectiveness. Regular updates based on new data and feedback from mechanics and customer service representatives ensured the models adapted to changing conditions and customer needs.
3. **Collaboration and Integration:** Successful implementation required close collaboration between data scientists, IT professionals, and automotive experts. Integrating the AI models with existing systems and workflows was key to achieving seamless operations and maximizing the benefits of predictive analytics.

Conclusion: AutoTech's journey of integrating AI into their automotive management systems highlights the transformative potential of predictive analytics in the automotive industry. By leveraging AI, the company not only addressed its vehicle maintenance and customer service challenges but also achieved significant improvements in operational efficiency and customer satisfaction. The case underscores the necessity of high-quality data, continuous model improvement, and effective integration for successful AI implementation. As a result, AutoTech set a new standard in the automotive industry for predictive maintenance and customer experience enhancement, positioning itself as a leader in innovative and reliable automotive services.

Implementing AI in Retail Banking for Customer Service Enhancement and Fraud Detection

Background: In the retail banking industry, providing excellent customer service and ensuring robust fraud detection are crucial for maintaining customer trust and satisfaction. BankSecure, a leading retail bank, was facing challenges with efficiently managing customer inquiries and detecting fraudulent activities. Traditional methods were insufficient in addressing the high volume of customer interactions and the sophisticated nature of modern fraud tactics. To tackle these issues, BankSecure decided to implement AI-powered solutions to enhance customer service and improve fraud detection.

Objective: The primary objective was to leverage AI to enhance customer service through intelligent chatbots and to improve fraud detection using predictive analytics. BankSecure aimed to reduce response time for customer inquiries by 50% and increase fraud detection accuracy by 35%.

Implementation:

1. **Data Collection and Preparation:** BankSecure began by collecting extensive data from their banking operations, including customer transaction histories, interaction logs, and known fraud patterns. Additional data sources included customer feedback and external financial data. This data was crucial for training both the customer service and fraud detection AI models.
2. **AI Model Development:** The bank collaborated with data scientists to develop two sets of machine learning models:
    - **Customer Service Chatbot Models:** Designed to handle customer inquiries efficiently, providing quick and accurate responses by analyzing previous interaction logs and feedback.
    - **Fraud Detection Models:** Capable of analyzing transaction data to identify patterns and anomalies indicative of fraudulent activities.
3. **Iterative Model Improvement:** Recognizing the importance of continuous improvement, BankSecure adopted an iterative approach to model development. The models were regularly updated with new data and feedback from customer service representatives and fraud analysts, refining their accuracy over time.
4. **Integration with Banking Systems:** The AI-powered models were integrated with the bank's existing customer service and transaction monitoring systems. This integration allowed for real-time customer interaction handling and fraud detection based on AI predictions. The system provided customer service representatives and fraud analysts with actionable insights, enabling proactive measures.

Outcome: The implementation of AI-powered solutions led to significant improvements in customer service and fraud detection at BankSecure:

- **Reduction in Customer Inquiry Response Time:** The AI-powered chatbots efficiently handled a large volume of customer inquiries, reducing response time by 50%. This improvement ensured quicker resolutions and enhanced customer satisfaction.

- **Increase in Fraud Detection Accuracy:** The AI system's advanced algorithms improved fraud detection accuracy by 35%. This enhancement helped BankSecure stay ahead of evolving fraud tactics and better protect its customers' assets.
- **Enhanced Operational Efficiency:** The automated system streamlined operations and reduced the manual workload for customer service representatives and fraud analysts. Efficient customer interaction handling and fraud detection ensured smoother operations and improved overall productivity.

Key Lessons:

1. **Importance of High-Quality Data:** The success of the AI implementation was heavily dependent on the quality and comprehensiveness of the data collected. Accurate and detailed data inputs were essential for reliable customer service and fraud detection.
2. **Iterative Model Improvement:** Continuous refinement of the AI models was crucial for maintaining their accuracy and effectiveness. Regular updates based on new data and feedback from customer service representatives and fraud analysts ensured the models adapted to changing conditions and needs.
3. **Collaboration and Integration:** Successful implementation required close collaboration between data scientists, IT professionals, and banking experts. Integrating the AI models with existing systems and workflows was key to achieving seamless operations and maximizing the benefits of AI solutions.

Conclusion: BankSecure's journey of integrating AI into their customer service and fraud detection systems highlights the transformative potential of AI in the retail banking industry. By leveraging AI, the bank not only addressed its customer service and fraud detection challenges but also achieved significant improvements in operational efficiency and customer satisfaction. The case underscores the necessity of high-quality data, continuous model improvement, and effective integration for successful AI implementation. As a result, BankSecure set a new standard in the retail banking industry for customer service enhancement and fraud detection, positioning itself as a leader in innovative and reliable banking services.

Implementing AI in Media and Entertainment for Content Personalization and Audience Engagement

Background: In the media and entertainment industry, providing personalized content and engaging audiences are crucial for maintaining viewership and increasing subscription rates. StreamMax, a leading media and entertainment company, was facing challenges with understanding viewer preferences and enhancing audience engagement. Traditional methods were insufficient in addressing the diverse content preferences of their global audience. To tackle these issues, StreamMax decided to implement AI-powered predictive analytics to enhance content personalization and audience engagement.

Objective: The primary objective was to leverage AI to provide personalized content recommendations and improve audience engagement. StreamMax aimed to increase viewer retention by 25% and boost engagement rates by 30%.

Implementation:

1. **Data Collection and Preparation:** StreamMax began by collecting extensive data on viewer interactions, including viewing histories, search queries, ratings, and demographic information. Additional data sources included social media activity and external content trends. This data was crucial for training the predictive analytics models.
2. **AI Model Development:** The company collaborated with data scientists to develop machine learning models capable of analyzing the collected data to predict viewer preferences and recommend personalized content. The models were designed to identify patterns and correlations that influence viewing behavior. Rigorous testing and validation ensured the models' accuracy and reliability.
3. **Iterative Model Improvement:** Recognizing the importance of continuous improvement, StreamMax adopted an iterative approach to model development. The predictive models were regularly updated with new data and feedback from viewers, refining their accuracy over time.
4. **Integration with Streaming Platform:** The AI-powered models were integrated with the company's existing streaming platform. This integration allowed for real-time personalization of content recommendations based on AI predictions. The system provided viewers with tailored content suggestions, enhancing their overall experience.

Outcome: The implementation of AI-powered predictive analytics led to significant improvements in content personalization and audience engagement at StreamMax:

- **Increase in Viewer Retention:** The AI system's accurate content recommendations resulted in a 25% increase in viewer retention. This improvement ensured that viewers stayed engaged with the platform for longer periods.
- **Boost in Engagement Rates:** By providing personalized content suggestions and enhancing the viewing experience, the company boosted engagement rates by 30%. Viewers interacted more with the platform, exploring recommended content and participating in discussions.

- **Enhanced Operational Efficiency:** The automated system streamlined operations and reduced the manual workload for content curators. Efficient content personalization ensured smoother operations and improved overall productivity.

Key Lessons:

1. **Importance of High-Quality Data:** The success of the AI implementation was heavily dependent on the quality and comprehensiveness of the data collected. Accurate and detailed data inputs were essential for reliable content recommendations and audience engagement.
2. **Iterative Model Improvement:** Continuous refinement of the predictive models was crucial for maintaining their accuracy and effectiveness. Regular updates based on new data and feedback from viewers ensured the models adapted to changing preferences and trends.
3. **Collaboration and Integration:** Successful implementation required close collaboration between data scientists, IT professionals, and content curators. Integrating the AI models with existing systems and workflows was key to achieving seamless operations and maximizing the benefits of predictive analytics.

Conclusion: StreamMax's journey of integrating AI into their streaming platform highlights the transformative potential of predictive analytics in the media and entertainment industry. By leveraging AI, the company not only addressed its content personalization and audience engagement challenges but also achieved significant improvements in operational efficiency and viewer satisfaction. The case underscores the necessity of high-quality data, continuous model improvement, and effective integration for successful AI implementation. As a result, StreamMax set a new standard in the media and entertainment industry for content personalization and audience engagement, positioning itself as a leader in innovative and engaging media services.

Implementing AI in the Public Sector for Citizen Services and Resource Optimization

Background: In the public sector, providing efficient citizen services and optimizing resource allocation are crucial for improving public satisfaction and operational efficiency. CityGov, a major municipal government, was facing challenges with managing citizen services and allocating resources effectively. Traditional methods were insufficient in addressing the diverse needs of citizens and the complexity of resource management. To tackle these issues, CityGov decided to implement AI-powered predictive analytics to enhance citizen services and optimize resource allocation.

Objective: The primary objective was to leverage AI to improve the efficiency of citizen services and optimize the allocation of municipal resources. CityGov aimed to reduce response times for citizen inquiries by 40% and improve resource utilization by 25%.

Implementation:

Data Collection and Preparation: CityGov began by collecting extensive data on citizen interactions, service requests, resource usage, and environmental conditions. Additional data sources included social media feedback and public surveys. This data was crucial for training the predictive analytics models.

AI Model Development: The municipal government collaborated with data scientists to develop machine learning models capable of analyzing the collected data to predict citizen needs and optimize resource allocation. The models were designed to identify patterns and correlations that influence service demand and resource usage. Rigorous testing and validation ensured the models' accuracy and reliability.

Iterative Model Improvement: Recognizing the importance of continuous improvement, CityGov adopted an iterative approach to model development. The predictive models were regularly updated with new data and feedback from public service employees, refining their accuracy over time.

Integration with Municipal Systems: The AI-powered models were integrated with the city's existing municipal management systems. This integration allowed for real-time analysis of service requests and automated resource allocation based on AI predictions. The system provided city officials and public service employees with actionable insights, enabling proactive measures to address citizen needs and optimize resource usage.

Outcome: The implementation of AI-powered predictive analytics led to significant improvements in citizen services and resource optimization at CityGov:
Reduction in Response Times for Citizen Inquiries: The AI system's ability to predict and prioritize service requests resulted in a 40% reduction in response times. This improvement ensured quicker resolutions and enhanced public satisfaction.

Improvement in Resource Utilization: By optimizing resource allocation based on AI recommendations, the city improved resource utilization by 25%. This enhancement led to cost savings and more efficient use of municipal assets.

Enhanced Operational Efficiency: The automated system streamlined operations and reduced the manual workload for public service employees. Efficient service request handling and resource allocation ensured smoother operations and improved overall productivity.

Key Lessons:

Importance of High-Quality Data: The success of the AI implementation was heavily dependent on the quality and comprehensiveness of the data collected. Accurate and detailed data inputs were essential for reliable service predictions and resource optimization.
Iterative Model Improvement: Continuous refinement of the predictive models was crucial for maintaining their accuracy and effectiveness. Regular updates based on new data and feedback from public service employees ensured the models adapted to changing conditions and citizen needs.

Collaboration and Integration: Successful implementation required close collaboration between data scientists, IT professionals, and public service experts. Integrating the AI models with existing systems and workflows was key to achieving seamless operations and maximizing the benefits of predictive analytics.

Conclusion: CityGov's journey of integrating AI into their municipal management systems highlights the transformative potential of predictive analytics in the public sector. By leveraging AI, the city not only addressed its service delivery and resource management challenges but also achieved significant improvements in operational efficiency and public satisfaction. The case underscores the necessity of high-quality data, continuous model improvement, and effective integration for successful AI implementation. As a result, CityGov set a new standard in the public sector for citizen services and resource optimization, positioning itself as a leader in innovative and efficient municipal governance.

Reference AI Tools

Automate Tasks and Operations

1. **Lavender.ai** https://lavender.ai/ is an AI-powered data entry tool that can help businesses save time and money by automating data entry tasks. Lavender.ai can extract data from various sources, including PDFs, spreadsheets, and images.
2. **MarketMuse** https://www.marketmuse.com/ is an AI-powered content marketing tool that can help businesses create more engaging and compelling content. MarketMuse uses AI to analyze content for grammar, spelling, and clarity issues. It also provides suggestions for how to improve the content.
3. **Phrasee** https://phrasee.co/ is an AI-powered copywriting tool that can help businesses create more persuasive and engaging marketing copy. Phrasee uses AI to generate text that is tailored to the specific audience that the business is trying to reach.
4. **Crystal Knows** https://www.crystalknows.com/ is an AI-powered customer service tool that can help businesses provide better customer service. Crystal Knows uses AI to understand customer intent and to generate responses that are tailored to the specific customer's needs.
5. **InsideSales** https://www.insidesales.com/ is an AI-powered sales automation tool that can help businesses close more deals. InsideSales uses AI to identify and qualify leads, schedule appointments, and track the progress of sales opportunities.
6. **Zia by Zoho CRM** https://www.zoho.com/crm/zia.html is an AI-powered CRM tool that can help businesses manage their customer relationships more effectively. Zia can automate tasks such as lead generation, customer support, and sales forecasting.
7. **Gemini** http://gemini.google.com/ is an AI-powered marketing automation tool that can help businesses generate more leads and sales. Sniper AI uses AI to identify and target potential customers, create personalized marketing campaigns, and track the results of marketing campaigns.
8. **ChatGPT** https://chat.openai.com/auth/login is an AI-powered chatbot that can help businesses provide customer service 24/7. ChatGPT can answer customer questions, resolve customer issues, and upsell and cross-sell products and services.
9. **DALL-E** https://openai.com/dall-e-3 is an AI-powered image generation tool that can help businesses create visually appealing marketing materials. DALL-E can generate images based on text descriptions, which can be used for product images, social media posts, and email campaigns.
10. **LiveChatAI** https://livechatai.com/ is an AI-powered live chat tool that can help businesses provide customer support in real time. LiveChatAI can answer customer questions, resolve customer issues, and upsell and cross-sell products and services.

Personalize Customer Experiences

1. **Chatfuel** https://chatfuel.com/ is a chatbot builder that allows businesses to create powerful chatbots for Facebook Messenger. Chatfuel chatbots can answer customer questions, provide product recommendations, and even complete transactions.
2. **Landbot** https://landbot.io/ is another chatbot builder that can be used to create chatbots for various channels, including websites, SMS, and email. Landbot chatbots are designed to help businesses generate leads, increase sales, and provide customer support.
3. **Flow.ai** https://app.flow.ai/onboarding is an AI-powered virtual assistant that can automate customer service tasks, such as answering questions, resolving issues, and scheduling appointments. Flow.ai can also be used to collect customer feedback and generate reports.
4. **Aira** https://aira.io/ is a mobile app that allows businesses to connect with a team of AI-powered virtual assistants who can help blind and low-vision individuals with customer service tasks, such as answering questions, providing product recommendations, and making travel arrangements.
5. **Amazon Lex** https://aws.amazon.com/lex/ is a cloud-based service that allows businesses to create conversational AI experiences using voice and text. Amazon Lex can be used to create chatbots, virtual assistants, and other applications that can naturally interact with customers.
6. **Google Dialogflow** https://cloud.google.com/dialogflow is a cloud-based platform that allows businesses to create conversational AI applications using natural language processing and machine learning. Google Dialogflow can be used to create chatbots, virtual assistants, and other applications that can naturally interact with customers.
7. **IBM Watson Assistant** https://www.ibm.com/products/watsonx-assistant is a cloud-based platform that allows businesses to create conversational AI applications using natural language processing and machine learning. IBM Watson Assistant can create chatbots, virtual assistants, and other applications that interact naturally with customers.
8. **Microsoft Bot Framework** https://dev.botframework.com/ is a cloud-based enterprise platform that allows businesses to create conversational AI applications using natural language processing and machine learning. Microsoft Bot Framework can create chatbots, virtual assistants, and other applications that naturally interact with customers.
9. **Pega Customer Decision Hub** https://www.pega.com/products/decision-hub is a cloud-based platform that allows businesses to create personalized customer experiences using AI and machine learning. Pega Customer Decision Hub can personalize marketing campaigns, customer service interactions, and customer touchpoints.
10. **Adobe Experience Platform** https://business.adobe.com/products/experience-platform/adobe-experience-platform.html is a cloud-based platform that allows businesses to collect, manage, and analyze customer data. Adobe Experience Platform can personalize customer experiences using AI and machine learning.

Automate Repetitive Tasks

1. **Zapier** https://zapier.com/apps/ai/integrations is a tool that helps you connect different apps and services to automate tasks between them. For example, you could use Zapier to automatically send an email to a customer when they purchase on your website or to automatically create a new lead in your CRM system when someone fills out a form on your website.
2. **Make** https://www.make.com/en is another tool that helps you automate tasks between different apps and services. Make is similar to Zapier, but it offers a broader range of features and integrations.
3. **Microsoft Power Automate** https://www.microsoft.com/en-us/power-platform/products/power-automate is a tool that helps you automate tasks within Microsoft Office and other Microsoft products.
4. **Google Cloud Platform** https://cloud.google.com/ offers a variety of AI-powered tools that can be used to automate tasks in businesses. For example, you could use Google Cloud Vision to classify images or extract text from documents automatically.
5. **Amazon Web Services** http://aws.amazon.com offers a variety of AI-powered tools that can be used to automate tasks in businesses. For example, you could use **Amazon Rekognition** to detect faces in images automatically or to transcribe audio recordings automatically.
6. **IBM Watson** https://www.ibm.com/products/watsonx-ai is a suite of AI-powered tools that can be used to automate tasks in businesses. For example, you could use Watson Assistant to create a chatbot that can answer customer questions or use Watson Discovery to find insights into your data.
7. **Salesforce Einstein** https://www.salesforce.com/products/einstein-ai-solutions/ is a suite of AI-powered tools that can be used to automate tasks in businesses. For example, you could use Einstein Prediction Builder to predict which customers will most likely churn or use **Einstein Engagement** Insights to see how customers interact with your website.
8. **Oracle AI Cloud** https://www.oracle.com/artificial-intelligence/ is a suite of AI-powered tools that can be used to automate tasks in businesses. For example, you could use **Oracle Data Science Cloud** to build machine learning models that can predict customer behavior or use **Oracle Analytics Cloud** to visualize your data.
9. **SAP Leonardo** https://www.sap.com/products/artificial-intelligence.html is a suite of AI-powered tools that can be used to automate tasks in businesses. For example, you could use **SAP Leonardo Machine Learning** to build machine learning models that can predict customer churn or use **SAP Leonardo Internet of Things** to collect data from your devices and use it to improve your business processes.
10. **Pegasystems** https://www.pega.com/ is a company that offers a variety of AI-powered tools that can be used to automate tasks in businesses like creating a chatbot that can answer customer questions or use **Pegasystems Customer Relationship Management** to manage customer interactions.

Manage Inventory

1. **Amazon Forecast** https://aws.amazon.com/forecast/ is a fully managed service that uses machine learning to forecast demand for your products. This can help you to avoid overstocking or understocking and to improve your inventory turns.
2. **Google AI Platform** https://cloud.google.com/ai-platform/docs/technical-overview offers various AI services that can be used to optimize inventory levels, including forecasting, demand planning, and inventory optimization.
3. **IBM Watson AI** https://www.ibm.com/products/watsonx-ai can be used to build custom AI models to optimize inventory levels for your specific business.
4. **Microsoft Azure Machine Learning** https://azure.microsoft.com/en-us/products/machine-learning offers various AI services that can be used to optimize inventory levels, including forecasting, demand planning, and inventory optimization.
5. **SAP Leonardo** https://www.sap.com/products/artificial-intelligence.html is a suite of AI-powered applications that can improve various business processes, including inventory management.
6. **Oracle Analytics Cloud** https://www.oracle.com/business-analytics/analytics-platform/ offers a variety of AI-powered analytics tools that can be used to optimize inventory levels, including forecasting, demand planning, and inventory optimization.
7. **Salesforce Einstein** https://www.oracle.com/business-analytics/analytics-platform/ is a suite of AI-powered applications that can improve various business processes, including inventory management.
8. **Shopify Inventory Management** https://www.shopify.com/retail/inventory-management is an app that uses AI to help businesses track inventory levels forecast demand and optimize their reorder points.
9. **Ecomdash Inventory Management** https://www.bigcommerce.com/apps/ecomdash/ is a cloud-based inventory management software that offers AL-powered features to help businesses optimize their inventory levels.

Manage Energy

1. **Opower** https://www.oracle.com/utilities/opower-energy-efficiency/ is a software company that uses AI to help utilities communicate with their customers about energy usage. Opower's software provides customers with personalized feedback on their energy usage, as well as tips on how to reduce their energy consumption.
2. **EnergyIQ** http://www.energyiq-usa.com/ is a software company that provides businesses with tools to help them monitor and manage their energy usage. EnergyIQ's software includes energy usage data visualization, energy efficiency reports, and energy savings plans.
3. **Enelx** https://www.enelx.com/n-a/en is a software company that provides businesses with a way to manage their energy usage through their demand response programs. EnerNOC's software allows firms to participate in demand response programs, which can help them reduce their energy costs.
4. **PowerFactor** https://powerfactors.com/ is a software company that provides businesses with a way to improve their energy efficiency through power factor optimization programs. PowerFactor's software helps businesses to reduce their energy consumption and improve their power quality.
5. **Enablon** https://www.wolterskluwer.com/ is a software company that provides businesses with tools to help them manage their environmental impact. Enablon's software includes energy usage data visualization, environmental impact reports, and environmental compliance tracking.
6. **SAP Greenit** https://www.sap.com/assetdetail/2022/08/38c39e9b-3c7e-0010-bca6-c68f7e60039b.html is a software solution from SAP that helps businesses manage their energy usage and environmental impact. SAP Greenit includes energy usage data visualization, environmental impact reports, and environmental compliance tracking.

Enhance Cybersecurity

1. **CrowdStrike Falcon** https://www.crowdstrike.com/falcon-platform/ is a cloud-native endpoint protection platform that uses AI to identify and respond to threats in real time. It can monitor traffic, detect anomalies, and block attacks before they cause damage.
2. **Darktrace** https://darktrace.com/ is an AI-powered threat detection platform that uses machine learning to identify and respond to threats across the entire IT estate. It can monitor traffic, detect anomalies, and block attacks before they cause damage.
3. **Cynet** https://www.cynet.com/ is an AI-powered security platform that provides visibility and control over all assets, users, and activities. It can monitor traffic, detect anomalies, and block attacks before they cause damage.
4. **FireEye Helix** https://www.threatprotectworks.com/Helix.asp is a cloud-based security platform that uses AI to detect and respond to threats across the entire IT estate. It can monitor traffic, detect anomalies, and block attacks before they cause damage.
5. **Check Point Infinity** https://www.checkpoint.com/infinity/ is a unified security platform that uses AI to protect against all types of threats, including malware, ransomware, and zero-day attacks. It can monitor traffic, detect anomalies, and block attacks before they cause damage.
6. **Symantec Endpoint Protection Enterprise** https://www.broadcom.com/products/cybersecurity/endpoint/end-user/enterprise is a cloud-based endpoint protection solution that uses AI to protect against malware, ransomware, and other threats. It can monitor traffic, detect anomalies, and block attacks before they cause damage.
7. **Sophos Intercept X** https://www.sophos.com/en-us/products/endpoint-antivirus is an endpoint protection solution that uses AI to detect and block threats. It can monitor traffic, detect anomalies, and block attacks before they cause damage.
8. **Fortinet FortiGate** https://login.forticloud.com/ is a next-generation firewall that uses AI to detect and block threats. It can monitor traffic, detect anomalies, and block attacks before they cause damage.
9. **Palo Alto Networks WildFire** https://www.paloaltonetworks.com/network-security/wildfire is a cloud-based malware analysis service that uses AI to detect and block malware. It can monitor traffic, detect anomalies, and block attacks before they cause damage.
10. **Vectra AI** https://www.vectra.ai/ is an AI-powered threat detection platform that uses machine learning to identify and respond to threats across the entire IT estate. It can monitor traffic, detect anomalies, and block attacks before they cause damage.

Detect Fraud

1. **Signifyd** https://www.signifyd.com/ is a fraud prevention platform that uses machine learning to identify and prevent fraudulent transactions. It offers a variety of features, including real-time fraud detection, chargeback protection, and risk scoring.
2. **Sift** https://sift.com/ is a fraud prevention platform that uses machine learning to identify and prevent fraudulent transactions. It offers a variety of features, including real-time fraud detection, risk scoring, and customer reviews.
3. **Kount** https://kount.com is a fraud prevention platform that uses a variety of data points, including device fingerprinting, IP addresses, and payment information, to identify and prevent fraudulent transactions. It offers a variety of features, including real-time fraud detection, risk scoring, and chargeback protection.
4. **DataDome** https://datadome.co/ is a fraud prevention platform that uses artificial intelligence to identify and prevent fraudulent websites. It offers a variety of features, including real-time website scanning, risk scoring, and blocking of fraudulent websites.
5. **NoFraud** https://www.nofraud.com/ is a fraud prevention platform that uses machine learning to identify and prevent fraudulent transactions. It offers a variety of features, including real-time fraud detection, risk scoring, and chargeback protection.
6. **FraudLabs Pro** https://www.fraudlabspro.com/ is a fraud prevention platform that uses a variety of data points, including IP addresses, credit card numbers, and email addresses, to identify and prevent fraudulent transactions. It offers a variety of features, including real-time fraud detection, risk scoring, and customer reviews.
7. **Subuno** https://www.subuno.com/ is a fraud prevention platform that uses machine learning to identify and prevent fraudulent transactions. It offers a variety of features, including real-time fraud detection, risk scoring, and chargeback protection.
8. **Covery** https://covery.ai/ is a fraud prevention platform that uses a variety of data points, including device fingerprinting, IP addresses, and payment information, to identify and prevent fraudulent transactions. It offers a variety of features, including real-time fraud detection, risk scoring, and chargeback protection.
9. **Vesta** https://www.vesta.io/ is a fraud prevention platform that uses artificial intelligence to identify and prevent fraudulent websites. It offers a variety of features, including real-time website scanning, risk scoring, and blocking of fraudulent websites.
10. **Cybersource** https://www.cybersource.com/en-us.html is a fraud prevention platform that offers a variety of features, including real-time fraud detection, risk scoring, and chargeback protection.

Make Decisions

1. **Tableau** https://www.tableau.com/ is a data visualization and analytics platform that helps businesses make better decisions by providing insights into their data. Tableau can be used to analyze data from a variety of sources, including sales data, customer data, and financial data. Tableau can also be used to create interactive dashboards and reports that can be shared with stakeholders.
2. **Domo** https://www.domo.com/ is a data visualization and analytics platform that helps businesses make better decisions. Domo is cloud-based, so it can be accessed from anywhere. Domo also has several features that make it easy to share data and insights with stakeholders.
3. **Sisense** https://www.sisense.com is a data analytics platform that helps businesses make better decisions by providing insights into their data. Sisense can be used to analyze data from a variety of sources, including sales data, customer data, and financial data. Sisense also has several features that make it easy to share data and insights with stakeholders.
4. **QlikView** https://www.qlik.com/us/products/qlikview is a data visualization and analytics platform that helps businesses make better decisions by providing insights into their data. QlikView can be used to analyze data from a variety of sources, including sales data, customer data, and financial data. QlikView also has several features that make it easy to share data and insights with stakeholders.
5. **MonkeyLearn** https://monkeylearn.com/ is an AI platform that helps businesses automate tasks and make better decisions. MonkeyLearn can be used to build machine learning models that can be used to classify data, extract insights, and automate tasks.
6. **RapidMiner** https://rapidminer.com/ is an open-source AI platform that helps businesses build and deploy machine learning models. RapidMiner can build models for various tasks, including classification, regression, and clustering.
7. **H2O.ai** https://h2o.ai/ is an open-source AI platform that helps businesses build and deploy machine learning models. H2O.ai can build models for various tasks, including classification, regression, and clustering.
8. **DataRobot** https://www.datarobot.com/ is a cloud-based AI platform that helps businesses build and deploy machine learning models. DataRobot can be used to build models for various tasks, including classification, regression, and clustering.
9. **IBM Watson** https://www.ibm.com/watsonx is a cloud-based AI platform that helps businesses build and deploy machine learning models. IBM Watson can be used to build models for various tasks, including classification, regression, and clustering.
10. **Google Cloud AI** https://cloud.google.com/products/ai Platform is a cloud-based AI platform that helps businesses build and deploy machine learning models. Google Cloud AI Platform can be used to build models for various tasks, including classification, regression, and clustering.

Forecast Sales

1. **HubSpot CRM** https://www.hubspot.com/products/crm is a customer relationship management (CRM) platform that helps businesses track leads, manage customer interactions, and close deals. It also includes AI-powered features that can help businesses predict future sales trends. For example, HubSpot's Sales Insights tool uses machine learning to analyze historical sales data and identify patterns that can be used to predict future sales.
2. **Zoho CRM** https://www.zoho.com/crm/ is a popular platform offering AI-powered sales forecasting features. Zoho's SalesIQ tool uses machine learning to analyze historical sales data and identify factors correlated with high sales. This information can then be used to predict which leads are most likely to convert into customers.
3. **Salesforce** https://www.salesforce.com/ is a leading CRM platform that offers a wide range of AI-powered features, including sales forecasting. Salesforce's Einstein Prediction Builder tool uses machine learning to analyze historical sales data and identify patterns that can be used to predict future sales.
4. **Pipedrive** https://www.pipedrive.com/ is a CRM platform designed specifically for businesses. Pipedrive's AI-powered Sales Booster tool uses machine learning to analyze historical sales data and identify patterns that can be used to predict future sales.
5. **Close.io** https://www.close.com/ is a CRM platform designed for sales teams that sell on a monthly or annual subscription basis. Close.io's AI-powered Sales AI tool uses machine learning to analyze historical sales data and identify patterns that can be used to predict future sales.
6. **Salesboom** https://admin.minicoursegenerator.com/mini-course/43524/Salesboom.ai is a CRM platform that is designed for sales teams that sell to businesses. Salesboom's AI-powered Sales insights tool uses machine learning to analyze historical sales data and identify patterns that can be used to predict future sales.
7. **Act-On** https://act-on.com/ is a marketing automation platform that offers AI-powered sales forecasting features. Act-On's Sales Intelligence tool uses machine learning to analyze historical sales data and identify factors correlated with high sales. This information can then be used to predict which leads are most likely to convert into customers.
8. **Marketo's Predictive Intelligence** https://business.adobe.com/products/marketo/adobe-marketo.html platform offers AI-powered sales forecasting features. Marketo's Predictive Intelligence tool uses machine learning to analyze historical sales data and identify patterns that can be used to predict future sales.
9. **Adobe Experience Cloud** https://business.adobe.com/ is a suite of marketing and sales tools that includes AI-powered sales forecasting features. Adobe's Sales Cloud uses machine learning to analyze historical sales data and identify patterns that can be used to predict future sales.
10. **Oracle Sales Cloud** https://www.oracle.com/cx/sales/ is a CRM platform that offers a wide range of AI-powered features, including sales forecasting. Oracle's Sales Cloud Einstein Forecasting tool uses machine learning to analyze historical sales data and identify patterns that can be used to predict future sales.

Predict Maintenance Operations

1. **Upkeep** https://upkeep.com/ is a cloud-based asset management platform that helps small businesses track the condition of their equipment and identify areas where maintenance is needed. Upkeep also offers predictive maintenance features that can help businesses avoid costly downtime.
2. **Fiix** https://fiixsoftware.com/ is a software-as-a-service (SaaS) platform that helps businesses manage maintenance operations. Fiix offers a variety of features, including predictive maintenance, asset tracking, and work order management.
3. **Asset Panda** https://www.assetpanda.com/ is a mobile-first asset management platform that helps businesses track equipment and inventory. Asset Panda also offers predictive maintenance features that help businesses avoid costly downtime.
4. **ServiceMax** https://www.servicemax.com/ platform helps businesses manage their field service operations. ServiceMax offers a variety of features, including predictive maintenance, asset tracking, and work order management.
5. **Oracle Field Service Cloud** https://www.oracle.com/cx/service/field-service-management/ is a field service management platform that helps businesses manage their field service operations. Oracle Field Service Cloud offers a variety of features, including predictive maintenance, asset tracking, and work order management.
6. **Microsoft Dynamics 365 Field Service** https://www.microsoft.com/en-us/dynamics-365 is a field service management platform that helps businesses manage their field service operations. It offers a variety of features, including predictive maintenance, asset tracking, and work order management.
7. **SAP Field Service Management** https://www.sap.com/products/scm/field-service-management.html is a platform that helps businesses manage their field service operations including predictive maintenance, asset tracking, and work order management.
8. **IBM Maximo Asset Management** https://www.ibm.com/products/maximo/asset-management is an asset management platform that helps businesses track their equipment and inventory

Glossary of Key AI Terms and Concepts

1. Algorithm: A set of instructions or rules that a computer follows to solve a problem or perform a task.

2. Artificial Intelligence (AI): The ability of a computer or machine to mimic the intelligence of humans. This includes learning from experience, understanding language, and making decisions.

3. Automation: The use of AI and machines to perform tasks without human intervention, often to increase efficiency and reduce errors.

4. Bias: When a machine learning model makes unfair or incorrect predictions because of prejudices in the training data.

5. Big Data: Large and complex datasets that traditional data processing tools cannot handle. AI and machine learning techniques are often used to analyze big data.

6. Chatbot: An AI program that can simulate conversation with users, often used for customer service and support.

7. Classification: A type of machine learning task where the goal is to assign labels or categories to data points. For example, sorting emails into "spam" or "not spam."

8. Computer Vision: A field of AI that enables computers to interpret and understand visual information from the world, such as images and videos.

9. Data: Information collected for analysis. In AI, data is used to train machines to recognize patterns and make decisions.

10. Deep Learning: A subset of machine learning that uses neural networks with many layers (deep networks) to analyze complex data and solve complicated problems.

11. Ethics in AI: The study of moral principles and guidelines for developing and using AI technologies in a way that is fair, responsible, and beneficial for society.

12. Feature: An individual measurable property or characteristic of the data used in machine learning. For example, age, height, and weight can be features.

13. Model: A mathematical representation created by a machine learning algorithm that makes predictions or decisions based on data.

14. Machine Learning (ML): A type of AI where computers learn from data and improve their performance over time without being explicitly programmed.

15. Natural Language Processing (NLP): A field of AI that focuses on the interaction between computers and humans through natural language. It involves understanding, interpreting, and generating human language.

16. Neural Network: A type of machine learning model designed to work like the human brain. It has layers of interconnected nodes (neurons) that process data.

17. Overfitting: When a machine learning model learns too much from the training data, including the noise, and performs poorly on new, unseen data.

18. Prediction: The result or outcome that a machine learning model generates based on the input data it receives.

19. Predictive Analytics: Using data, machine learning, and statistical techniques to make predictions about future events or behaviors.

20. Regression: A type of machine learning task where the goal is to predict a continuous value, like predicting the price of a house based on its features.

21. Reinforcement Learning: A type of machine learning where an agent learns to make decisions by taking actions in an environment to maximize a reward.

22. Supervised Learning: A type of machine learning where the model is trained on labeled data, meaning the input data is paired with the correct output.

23. Training Data: The data used to teach a machine learning model. It includes examples and outcomes that the model learns from.

24. Underfitting: When a machine learning model is too simple and cannot capture the patterns in the training data, leading to poor performance.

25. Unsupervised Learning: A type of machine learning where the model is trained on unlabeled data and must find patterns and relationships in the data on its own.

Made in United States
Orlando, FL
04 November 2026

90200175R00092